American Institute of Chemical Engineers

Bionanotechnology Plenary Sessions

Topical Conference at the
2007 AIChE Annual Meeting

November 4-9, 2007
Salt Lake City, Utah, USA

Printed from e-media with permission by:

Curran Associates, Inc.
57 Morehouse Lane
Red Hook, NY 12571
www.proceedings.com

ISBN: 978-1-60423-836-5

ISBN: 978-1-60423-836-5

American Institute of Chemical Engineers
Proceedings
Three Park Avenue
New York, NY 10016-5991
Phone: 212-591-8100

www.aiche.org

American Institute of Chemical Engineers

Bionanotechnology Plenary Sessions
2007

TABLE OF CONTENTS

14c Microarray Diagnostic Formats: Meaning, Metrics and Money

David W. Grainger, Department of Bioengineering and Department of Pharmaceutics and Pharmaceutical Chemistry, University of Utah, 301 Skaggs Hall, 20 S 2000 E, Salt Lake City, UT 84112

The capture of genetic material (DNA and RNA) from various life science samples – from food to microbiology to environment to medicine – is very important in identifying disease, genetics, and bio-hazards. DNA diagnostics constitute the class of techniques used. DNA microarrays are one recently developed miniaturized diagnostic device where thousands of DNA samples can be identified simultaneously with small sample size and high throughput. The method involves capture of one strand of DNA target by its complementary DNA probe partner on a surface. Captured DNA signal is then measured with optical or radiometric methods. We study how DNA and proteins interact with device capture surfaces in microarray diagnostic assays, and have used many methods to understand how to optimize surface capture and diagnostic assay performance. Multiple surface analysis methods include x-ray photoelectron spectroscopy (XPS), surface optical methods (SPR), 32P-radiolabeling, and fluorescence assays on surfaces of different chemistries. Higher DNA probe surface density does not improve partner DNA signal capture. Immobilized DNA for this duplex capture produces steric and electrostatic interference on surfaces that blocks DNA target analyte capture. But, reduced DNA probe density, allowing higher hybridization efficiencies, produces lower overall assay signal. DNA assay performance depends on understanding these effects in microarray microspot devices. Recent work from the Grainger group in this regard: P. Gong, G.M. Harbers, D.W. Grainger, "Multi-technique Comparison of Immobilized and Hybridized Oligonucleotide Surface Density on Commercial Amine-Reactive Microarray Slides," Analy. Chem. 78:7, 2342-2351 (2006). C-Y Lee, P. Gong, L.J. Gamble, D.G. Castner, D.W. Grainger, "Hybridization Behavior of Mixed DNA/Alkylthiol Monolayers on Gold: Characterization by SPR and 32P Radiometric Assay," Analy. Chem., 78, 3316-3325 (2006). P. Gong, C-Y Lee, G.M. Harbers, D.G. Castner, D.W. Grainger, L.J. Gamble, "Comparison of Immobilized DNA Orientation and Density Using NEXAFS, XPS, and Fluorescence Intensity Measurements on Gold" Analy. Chem., 78, 3326-3334 (2006) C-Y Lee, L.J. Gamble, D.W. Grainger, D.G. Castner, "Mixed DNA/Oligoethylene glycol functionalized gold surfaces improve DNA hybridization in complex media," BioInterphases, 1, 83-92 2006. P. Wu, D.W. Grainger, "Comparison of hydroxylated print additives on antibody microarray performance," J. Proteome Res., 5(11) 2956-2965, 2006. D. Dandy, P. Wu, D.W. Grainger, "Assay feature size influences nucleic acid capture in DNA microarrays," Proc Natl Acad Sci U S A 2007 104(20):8223-8 D. Xue, P. Gong, D.W. Grainger, C.M. Elliott, V. Disette, C.A. Bignozzi, S. Caramori, "Electrochemical detection of DNA hybridization through catalytic oxidation of Co(II) complexes by covalently bound or intercalated Ru(II) complexes," J Am Chem Soc 2007 129 (7):1854-5 C-Y Lee, Phuong-Cac T. Nguyen,1,2 D.W. Grainger, L.J. Gamble, D.G. Castner, Structure and DNA Hybridization Properties of Mixed Nucleic Acid/Maleimide-ethylene glycol Monolayers, Anal Chem 2007 Jun 15;79(12):4390-4400 C.Y. Lee, G. M. Harbers, D.W. Grainger, L. J. Gamble, D. G. Castner, "Fluorescence, XPS and ToF-SIMS Surface Chemical State Image Analysis of DNA Microarrays, J Am Chem Soc 2007 published on line Jul 11;

69a Polyelectrolyte Multilayers for Tunable Delivery of Antibiotics

Helen F. Chuang and Paula T. Hammond. Chemical Engineering, MIT, Room 66-525, 77 Massachusetts Avenue, Cambridge, MA 02139

Multilayered polymer thin films were constructed via the layer-by-layer (LbL) deposition technique using a unique water-hydrolyzable polycation, resulting in coatings that are stable in air but erode top-down in a layer-by-layer fashion when exposed to aqueous physiological environment. The multilayered nature of these films allows for the encapsulations and subsequent sequential releases of multiple drugs, with controls over the dosage and release kinetics of each encapsulated agent. We focus on the development of a coating for sequential delivery of pain killers with anti-inflammatories, antibiotics, and growth factors to be applied onto orthopedic implant surfaces. We have incorporated small-molecule antibiotics into these films without the need for any pre-modification, and demonstrated controls over encapsulation dosage and subsequent release rate through a variety of tuning factors. Several classes of antibiotics are under investigation, some with sustained release over one to two weeks, an ideal duration for infection control over biomedical implant surfaces. Aside from antimicrobial coatings, our findings can also be applied towards coating-based delivery of other therapeutic agents, including proteins and nucleic acids, as well as other active ingredients in the non-medical industries.

Web Page: web.mit.edu/hammond/lab/chuang.htm

69b Multilayered Films Fabricated From Charge-Shifting Cationic Polymers: Controlled Erosion And Sequential Release Of DNA From Surfaces

Xianghui Liu and David M. Lynn. Chemical and Biological Engineering, University of Wisconsin - Madison, 1415 Engineering Drive, Madison, WI 53706

Cationic polymers that assemble reversibly with anionic polymers under physiological conditions are of interest for a broad range of biotechnical and pharmaceutical applications, including gene and drug delivery and the development of new polyelectrolyte-based surface coatings. We have designed a new class of "charge-shifting" cationic polymers that permit control over the assembly and disassembly of macromolecular aggregates and assemblies in aqueous environments. The conjugation of ester-functionalized side chains to cationic polymer backbones yields polymers that change charge states dynamically via side chain hydrolysis when exposed to physiological media. These changes in net charge and charge density result in changes in the nature of electrostatic interactions within polycation/polyanion complexes and assemblies. For example, we have demonstrated that charge-shifting cationic polymers based on the conjugate addition of methyl acrylate to linear poly(ethylenimine) promote the reversible assembly of these polymers with DNA in physiologically relevant media and enhance the delivery of DNA to cells.

This presentation will focus on the application of these charge shifting materials to the layer-by-layer assembly of ultrathin, erodible multilayered polyelectrolyte films (e.g., from 40 nm to 100 nm thick) using model anionic polymers or plasmid DNA. These ultrathin films erode and release DNA when incubated in phosphate-buffered saline (PBS) at rates dependent on polymer structure, and we demonstrate that the DNA released from these films remains transcriptionally active and able to promote transgene expression in mammalian cells. We demonstrate further that these charge-shifting cationic polymers can be used to fabricate ultrathin films that erode in a controlled manner and permit the design of films that can be used to release two types of DNA sequentially (i.e., rapid release of a first type of DNA followed by slower release of a second DNA construct) and that the order in which these two DNA constructs are released can be dictated by the order in which they are deposited during film fabrication. The application of these erodible films to the promotion of surface-mediated delivery of DNA to cells will be discussed.

69c Uni-Nanogel For Controlled Delivery Of Multiple Therapeutics

Gauri Misra, Surgery, Penn State University College of Medicine, 500 University Drive, Hershey, PA 17033, William Weiss, Surgery, MC H151, Penn State College of Medicine, 500 University Drive, Hershey, PA 17033, and Tao Lu Lowe, Surgery, Pennsylvania State University, 500 University Drive, Hershey, PA 17033.

Nanotechnology has emerged as an exciting area of science with a wide range of applications in biology, medicine, and pharmaceutics[1-4]. Polymer-based nanogels are of especial interest in drug delivery and diagnostic applications because of tunable chemical and physical properties. In this work, we report the synthesis, characterization, and drug releasing property of nanogels based on a monomer, N-isopropyl acrylamide (NIPAAM) or 2-hydroxyethylmethacrylate and a cross-linkable macromer, 2-hydroxyethylmethacrylate-lactate-dextran. The nanogels with varying mass ratios between the monomer and macromer were synthesized in aqueous medium using ultra-violet photo-cross-linking and characterized by dynamic light scattering method. The nanogels were found to be non-toxic to neuron-like PC 12 and R28 cells, and Blood Brain Barrier Vascular Endothelial Cells (BBBVECs) up to 500 ìg/mL concentration. Blood biocompatibility of the nanogels was evaluated by hemolysis test. The nanogels were non-hemolytic for at least 24 hr. Model therapeutic agents, such as heparin, low molecular weight heparin (LMWH), insulin, and FITC-dextran having molecular weights 15,000, 4,500, 5,600, and 4,000, respectively, were loaded into the nanogels during the synthesis process. Release of these therapeutic agents in PBS (pH 7.4) at 37 °C occurred at different time and onset of the release depended on the size and conformation of the therapeutic agents. Insulin was released faster than LMWH and FITC-Dextran. Almost no release of heparin was observed for four months, and only HEMA-macromer nanogel released heparin after eighty days. Thus, the developed nanogels have great potential to be used as nanodevices for controlling the release of multiple drugs. Further work on encapsulating multiple drugs into and releasing them from the nanogels is in progress.

1. P. Ulanksi and J.M. Rosiak (2004). Polymeric Nano/Microgels. In "Encyclopedia of Nanoscience and Nanotechnology" (H. S. Nalwa, ed.), Vol. 8, pp. 845-871. American Scientific Publishers. 2. Peppas, N. A. (2004). Intelligent therapeutics: biomimetic systems and nanotechnology in drug delivery. Advanced Drug Delivery Reviews 56, 1529. 3. S. Nayak S, and L.A. Lyon, Soft nanotechnology with soft nanoparticles, Angew. Chem. Int. Ed. Engl., 44 (2005) 7686. 4. F. Chellat, Y. Merhi, A. Moreau, and L. Yahia, Therapeutic potential of nanoparticulate systems for macrophage targeting, Biomaterials, 26 (2005) 7260.

69d Doxorubicin Delivery Using Superparamagnetic Iron Oxides Nanoparticles Encapsulated Thermosensitive Liposome

Jae-Ho Lee and Joseph Frank. Experimental Neuroimaging Section, National Institutes of Health, Clinical Center NIH Building 10 Room B1N256, Bethesda, MD 20892-1074

Encapsulation of dextran coated superparamagentic iron oxide nanoparticles (SPION) and chemotherapeutic agents or cytokines inside various types of liposomes holds the potential for monitoring drug delivery by MRI and therapy combined with hyperthermia. SPION are a class of MRI contrast agent and depending on the crystal size of the iron oxide they can also be used as hyperperthermia agents via excitation by an alternative magnetic field (AMF).

We have synthesized stable liposomes that contain both SPION and doxorubicin (DOX), which allows for tracking by MRI and combines hyperthermia and chemotherapy treatment of cancer. The highly negatively charge of dextran coated SPION along with the positive charge on DOX hinders efficient encapsulation in liposomes at the same time. We have developed a successful strategy that encapsulates both dextran coated SPION and DOX in a thermosensitive liposome formulation at the same time. Our studies show that the encapsulation of DOX and dextran coated SPION results in complexation of the two agents followed by precipitation in solution. In addition, the current pH gradient method for maximizing doxorubicin encapsulation causes release of iron from SPION. To prevent the precipitation of the two agents, maximize doxorubicin encapsulation and minimize free iron we used polymer-coated techniques. This polymer coating approach gives not only stability through steric hindrance but also prevents free ironization. Our results indicate successful encapsulation and long term stability of SPION +DOX complex in liposomes (Lip-SPION+DOX) in culture media.

We will also present in vitro experiments using various tumor cell lines to determine the effect of the LIP-SPION+DOX agent on cell viability as a basis for in vivo studies.

69f Surface-Mediated Gene Delivery Of Peptide-Lipoplexes

Jennifer C. Rea, Annelise E. Barron, and Lonnie D. Shea. Chemical & Biological Engineering, Northwestern University, 2145 Sheridan Road, Tech E136, Evanston, IL 60208

The development of safe and efficient gene delivery methods is a fundamental goal of gene therapy. Delivery of vectors from biomaterials has been employed as a means to overcome mass transport limitations, which is particularly important for lipoplexes for which cellular internalization is a rate-limiting step. In this report, we investigate the hypothesis that the incorporation of cationic peptides into lipoplexes can modulate the interactions of the vector with the substrate, and may enhance cellular binding and internalization. Peptide-lipoplexes were formed by the addition of short, cationic peptides to DNA prior to the addition of Lipofectamine. Complexes were immobilized to tissue-culture polystyrene prior to the seeding of NIH3T3 cells. Surface-mediated transfection with peptide-lipoplexes resulted in a 7.4-fold increase in transfection efficiency compared to using lipids alone. In addition, the amount of peptide needed to enhance surface-mediated lipofection is less than the amount of peptide needed to enhance lipofection performed by bolus delivery of complexes. In addition to tissue-culture polystyrene, multiple surface chemistries are being investigated for surface-mediated gene delivery of peptide-lipoplexes. These peptide-lipoplexes demonstrate a robust and tunable gene delivery system that can be used to investigate the mechanisms of non-viral gene delivery.

69e Liposomal Delivery Of Ac-225 Targeted To Ovarian Cancer Cells

69g Multifunctional Lipopolyplex for Delivery of Antisense Oligonucleotide

Chee Guan Koh[1], Xulang Zhang[2], Bo Yu[1], Xiaojun Yang[3], Yan Jin[2], Jingjiao Guan[2], Zhengzheng Fei[1], Robert J. Lee[3], and L. James Lee[4]. (1) Department of Chemical and Biomolecular Engineering, The Ohio State University, 140 West 19th Avenue, Columbus, OH 43210, (2) NSF Center for Affordable Nanoengineering of Polymeric Biomedical Devices, The Ohio State University, 140 West 19th Avenue, Columbus, OH 43210, (3) College of Pharmacy, The Ohio State University, 542 Parks Hall, 500 West 12th Avenue, Columbus, OH 43210, (4) Chemical and Biomolecular Engineering, The Ohio State University, 125A Koffolt, 140 West 19th, Columbus, OH 43210

In recent years, the delivery of therapeutic genes by non-viral vectors (polyplexes, lipoplexes, or lipopolyplexes) has gained significant interest. The systemic application of non-viral vectors has been limited by rapid clearance from the bloodstream following intravenous injection. Non-viral vectors that contained surface shielding and targeting ligands such as poly(ethylene glycol) (PEG) and transferrin or folate, respectively, have been shown to improve circulation time and specificity.

In this study, we have developed multifunctional lipopolyplex containing antisense oligonucleotide for targeting Bcl-2 down regulation. Using K-562 human erythroleukemia cells as a model cell line, cell viability, gene transfection efficiency, and Bcl-2 down regulation are examined. We have characterized the lipopolyplex size and surface charge by dynamic light scattering and zeta potential measurement. In addition, a novel microfluidic approach will be used to synthesis the lipopolyplexes and compared with conventional bulk mixing method.

101a Development Of A Nanoporous Polymer Membrane For Implantable Drug Delivery Devices

Eric E. Nuxoll, Department of Pharmaceutics, University of Minnesota, 9-177 Weaver Densford Hall, 308 Harvard St. SE, Minneapolis, MN 55455, Marc A. Hillmyer, Dept. of Chemistry, University of Minnesota, 207 Pleasant St. SE, Minneapolis, MN 55455, and Ronald A. Siegel, Departments of Pharmaceutics and Biomedical Engineering, University of Minnesota, 9-177 Weaver Densford Hall, 308 Harvard St. SE, Minneapolis, MN 55455.

A common challenge for many proposed drug delivery systems is protection of the drug depot from antibodies and serum proteins. Unlike most current implantable devices, drug delivery devices require an open interface with the body through which to deliver pharmaceutical agents. How does one close off this interface to immunoproteins while permitting free transport of target solutes? In most cases, the large nature of the invading proteins makes size exclusion an attractive barrier option. To this end, we are adapting specialized triblock terpolymers for use as protective membranes in drug delivery/detection devices. We combine the nanoscale pore size of these polymers with thin, rigid silicon supports built using MEMS fabrication technologies.

We begin by coating a 100 μm thick silicon wafer with low-stress silicon nitride (LSN) using low-pressure CVD. Squares with 20 μm sides are then patterned on the bottom side by photolithography, etched into the LSN using reactive ion etching, and etched through the underlying silicon using Bosch-process directed reactive ion etching. The LSN on the top side of the wafer acts as an etch-stop, leaving a silicon membrane with uniform, straight microscale pores capped by a 100 nm layer of LSN.

A poly(styrene)-poly(isoprene)-poly(lactide) (PS-PI-PLA) block copolymer is then spin-coated onto this LSN surface and annealed under vacuum. The PLA block self-assembles into cylinders perpendicular to the coating, nearly spanning it. The PLA block is then etched away, leaving 30 nm pores in the polymer film. The device is then subjected to hydrofluoric acid vapor to remove the LSN capping the microscale pores and a final, brief oxygen plasma etch removes any PS capping the nanoscale pores.

The resulting composite membrane consists of a 100 nm thick PS layer with 30 nm wide pores overlaying a 100 μm thick silicon support with 20 μm wide pores. The nanoscale pores are lined with PI, offering a convenient handle for grafting other materials onto the pore walls, from simple immuno-masking agents like poly(ethylene glycol) to stimuli-responsive polymers for additional functionality. Such membranes should provide immuno-isolation without retarding small molecule transport and should integrate well with the burgeoning number of BioMEMS devices under development

101b Development Of A Microfluidic Hydrogel-Based Negative Pressure Pump

Tobias D. Wheeler and Abraham D. Stroock. Chemical and Biomolecular Engineering, Cornell University, 120 Olin Hall, Ithaca, NY 14853

A commonplace and yet remarkable phenomenon is the transport of water through trees: liquid water travels from sub-saturated soil, across the membranes of roots, and through meters of micron-scale ducts (xylem) in the trunk, branches and leaves, in order to maintain hydration of the tree as water evaporates from the leaves during CO2/O2 exchange with the atmosphere. This transport of water is termed transpiration. To travel from the roots of a tree to the highest leaves, liquid water that is initially at atmospheric pressure must be pulled across hydrodynamic resistances imposed by the 30-100 micron xylem ducts and against gravity. This requires the water to be placed at 10's of bars of negative pressure. In the plant science community, it is hypothesized that capillary pressures generated in nano-scale pores within the leaf allow trees to generate the required high negative pressures. We have developed a microfluidic pump (the synthetic tree) that instead relies on osmotic gradients across a hydrogel material to transduce the chemical potential driving force behind transpiration. I will present the thermodynamic, solid mechanical and transport considerations for the hydrogel material that acts as a pump in this system. I will then present experimental results that illustrate the success of the design as measured by the ability of the synthetic tree to use negative pressures to drive the transport of liquid water. Finally, I will discuss what further developments are required to improve the pumping ability of the synthetic tree and make it a viable contributor to other wick-based technologies, such as heat pipes and water management in air-breathing electrode fuel cells.

101c Post Bonding Modification of Microfluidic Channel Materials for the Manufacture of Bioanalytical Devices

Zhan Gao[1], Chang-Soo Kim[2], and David B. Henthorn[1]. (1) Department of Chemical and Biological Engineering, University of Missouri-Rolla, 143 Schrenk Hall, Rolla, MO 65409, (2) Department of Electrical and Computer Engineering, Dept. of Biological Sciences, University of Missouri-Rolla, Rolla, MO 65409

There has been increased interest in the use of negative-tone photoresist materials as structural components in microfluidic devices. However, the traditional steps used to seal a microchannel (heat, pressure, etc.) may be incompatible with biomolecular components often used in the development of bioanalytical devices. In this work, we have demonstrated a technique for post bonding functionalization of the photoresist SU-8 using thin polymer films. First, a photoactive initiator is conjugated on the SU-8 surface using a hydrogen abstraction technique. The now photoactive surface is then brought in contact with various monomeric species, with UV light used to induce a surface initiated polymerization reaction. Wettability of the normally hydrophobic SU-8 material was markedly enhanced by grafting hydrogels comprised of either 2-hydroxyethylmethacrylate (HEMA) monomer or macromonomers of poly(ethylene glycol) (PEG). The static water contact angle of freshly prepared SU-8 was reduced from 78° to 35° using grafted pHEMA hydrogels, and further reduced to 8° using PEG-rich hydrogels. Surface bound hydrogel layers containing the enzyme glucose oxidase were prepared to evaluate the applicability of the technique to integration of biomolecular components. In addition, incorporation of analyte-sensitive fluorophore layers was also investigated.

101d Tailored Recognitive Films Via Engineered Macromolecular Structure

Asa D. Vaughan, K. RyAnne Noss, and Mark E. Byrne. Biomimetic & Biohybrid Materials, Biomedical Devices, and Drug Delivery Laboratories, Department of Chemical Engineering, Auburn University, Auburn, AL 36849-5127

This work involves the creation and optimization of polymer thin films as enabling technology for robust point-of-care diagnostic platforms and drug delivery carriers. Recognitive polymeric networks can be rationally designed and produced through tailoring non-covalent complexation interactions between "guest" template molecules and functional monomers to produce polymer films with 'macromolecular memory.' The resulting networks can be highly selective to the template molecule or control release due to the particular chemistry and orientation of the polymer chains. In this work, structural characteristics of polymer films were altered and template binding/transport properties were analyzed. Specifically, crosslinking monomer length and concentration as well as free-radical initiation scheme were changed resulting in different polymer structures with various mesh sizes, polymer volume fractions, and template binding characteristics such as affinity, capacity, selectivity, and diffusion. Methacrylic acid-co-poly ethylene glycol dimethacrylate (poly(MAA-co-PEGDMA)) copolymer recognitive networks were prepared via UV-photopolymerization with template molecules such as ethyl adenine-9-acetate and testosterone. Results demonstrate that a living/controlled initiation scheme significantly increases the affinity and capacity of recognitive polymers. This is hypothesized to be due to a more uniform and homogeneous structural architecture. Additionally, a 77% increase in crosslinking monomer length resulted in polymers with a 6 fold decrease in affinity, a 2 fold decrease in capacity, and decreased selectivity with increased template diffusional transport. Similarly, increasing the crosslinking monomer concentration from 50 to 90% yielded polymers with a 4 fold increase in binding affinity, a 1.5 fold increase in capacity, as well as increased selectivity with decreased template transport. Engineering of templated polymer structural properties is vital to the further development of novel platform technologies that will utilize recognitive mechanisms and structures. Increasing the tailorability of recognitive biomaterials yields promising materials for robust, inexpensive, future point-of- care diagnostic and therapeutic devices.

Surface Molecularly Imprinted Polymers for the Recognition of Protein

Youyou Zheng, David B. Henthorn
Department of Chemical and Biological Engineering, University of Missouri – Rolla
Rolla, MO, 65409

Abstract

A surface molecular imprinting technique used for the production of protein recognitive materials is presented. Functional monomers were photopolymerized on a template mask with immunoglobulin immobilized in a specific molecular orientation. After polymerization, the molecularly imprinted materials were removed from the template mask, yielding a surface with memory for the template immunoglobulin. These materials were then tested for recognition of the desired molecule in a solution containing various competitive biomolecules. It was shown that the materials made with the surface imprinting technique were able to specifically recognize and bind the target molecule while rejecting other competing species. Easy preparation of the surface imprinted polymers, their high stability, and their selectivity toward biomacromolecules make this technique promising in the application of biological assays and sensors, etc..

1. Introduction

Due to their ability to specifically recognize and bind to a biomolecule out of a complex mixture of closely related species, antibodies are effective analytical reagents applied in immunoassays, immunoaffinity chromatography and immunosensors (1-7). However, the cost of producing natural antibodies is relatively high and the molecules are unable to sustain high temperatures, large shifts in pH, or exposure to organic solvents (8-9). Therefore, the design and synthesis of materials capable of mimicking natural antibodies is of obvious importance (10). One technique, molecular imprinting, has been increasingly adopted because of its ability to produce biomimetic receptor systems similar to their natural counterparts (11-12). The procedure of molecular imprinting usually includes the polymerization of functional monomers in the presence of a target molecule. After the polymerization, the target molecule is removed and the molecular cavity left inside the polymer networks thus possesses a structure complementary to the print molecule. As a result, the imprinted polymer is capable of recognizing and binding the target molecule (13). Molecularly imprinted polymers (MIPs) have been utilized primarily in conjunction with small molecules such as organic compounds, amino acids and their derivatives (14).

The imprinting of macromolecules like proteins has been studied, but is still considered difficult (15-19). Two negative effects frustrate the successful imprinting of macromolecules, steric hindrance and the thermodynamic effect (20). Diffusion of macromolecules in or out of polymer networks is hindered. Also, the shape of many macromolecules is sometimes non-rigid, resulting in heterogeneity in the imprinted binding sites.

In this paper, we discuss a surface molecular imprinting technique which has demonstrated success in the formation of recognitive materials targeted towards biomacromolecules. In the experiments, a monoclonal IgG immunoglobulin has been used as the templating molecule because of the unique conserved/variable structure. The conserved Fc domain of the immunoglobulin has been immobilized on an inert surface, while the Fab domain (site of antigen binding) of the immunoglobulin is facing upward. Due to this structure, we are able to investigate the ability to imprint one specific area (Fab domain) and compare with molecules which are structurally similar. Other advantages of the surface imprinting technique include the ease of separating the template from the newly formed MIP, and the possible ability to reuse the template mask. The scheme of experimental procedure is shown in Figure 1.

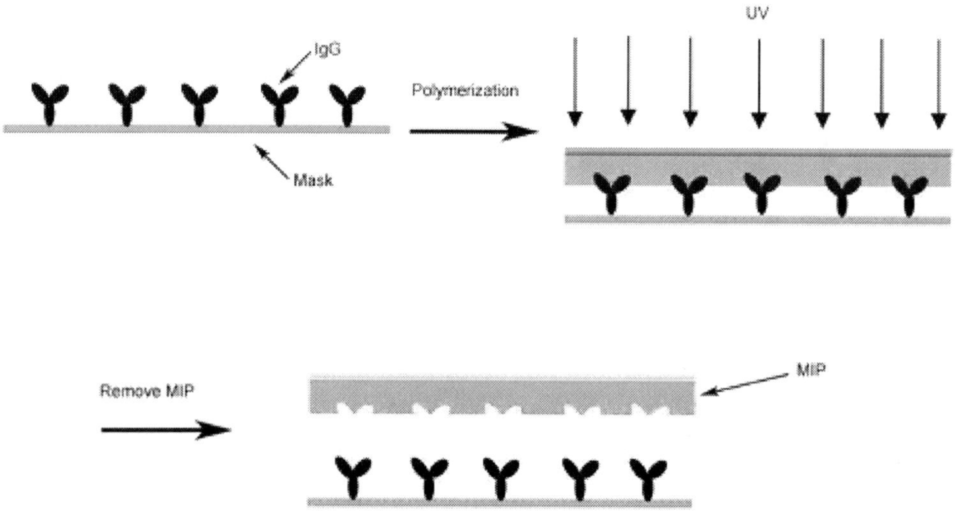

Figure 1: Surface imprinting procedure where polymerization is done in the presence of a templating mask with immobilized target molecule

2. Materials and Methods

2.1 Materials

Rabbit anti-pig IgG (whole molecule) alkaline phosphatase conjugate (RAP IgG-AP), monoclonal anti-c-myc purified mouse IgG (anti-c-myc IgG), monoclonal anti-c-myc alkaline phosphatase conjugate (anti-c-myc-AP), ethylene glycol dimethacrylate (EGDMA), bovine serum albumin (BSA) and 1-hydroxycyclohexyl phenyl ketone (HCPK) were purchased from Sigma-Aldrich (St. Louis, MO.). Sodium-meta-periodate ($NaIO_4$) and 2-hydroxyethylmethacrylate (HEMA) were from ACROS Organics. The functional monomer HEMA was vacuum distilled before use to remove any free radical inhibitor. Blank polystyrene microplates (Costar 3370), carbohydrate binding surface 96-well microplates (Corning 2507), and non-binding surface microplates (Corning 3990) were used. 1-step PNPP (p-Nitrophenyl Phosphate, Disodium Salt) and serum-derived IgG from a rabbit source

(Rabbit IgG whole molecule) were obtained from Pierce Biotechnology, Inc. Dimethyl sulfoxide (DMSO) was purchased from J. T. Baker.

2.2 Preparation of templating mask

In this work, the template molecule immunoglobulin was always immobilized on the carbohydrate binding microplates utilizing a hydrazide chemistry. First, the Fc domain of anti-c-myc IgG was oxidized by $NaIO_4$ following the procedure of antibody conjugation (21). Next, the functionalized IgG was reacted with the hydrazide coated surface of the Corning carbohydrate binding microplates. The concentration of anti-c-myc IgG added into each microwell was 0.001 mg/ml. Later, a blocking buffer, 1 wt. % BSA in PBS-Tween 20 buffer at pH 7.4, was used. After the blocking process, the now coated microplates were washed by the same PBS-Tween 20 buffer to remove any unbound protein.

2.3 Formation of control and molecularly imprinted polymers

Polymer discs were synthesized from a pre-polymerization mixture comprised of 20 mole % HEMA as functional monomer and 80 mole % EGDMA as crosslinker. The photoinitiator HCPK was then added to a concentration of 2 wt. %. In a nitrogen filled glove box, 45 μl of the pre-polymerization solution (HEMA, EGDMA and HCPK) and 30μl of the solvent DMSO were pipetted into the microwells. Free radical polymerization was initiated with UV light at 375 nm (UV Shark LED, Optotech, Inc. Wheeling, IL). The total polymerization time was 3 minutes with the intensity of 300 mW/cm^2 and the distance between the UV light source and pre-polymerization mixture was about 3 cm. The procedure is shown in Figure 2.

Control materials were synthesized in the same manner, except blank polystyrene microplate wells were used as the imprinting surface. In addition BSA coated surfaces were prepared to test the imprinting of an unrelated protein. To make BSA coated microplate templating masks, a solution containing 1 wt. % BSA in PBS-Tween 20 buffer (pH 7.4) was added into a blank polystyrene microplate for 15 minutes at room temperature. The microplate wells were then washed with the same buffer three times to remove unbound protein.

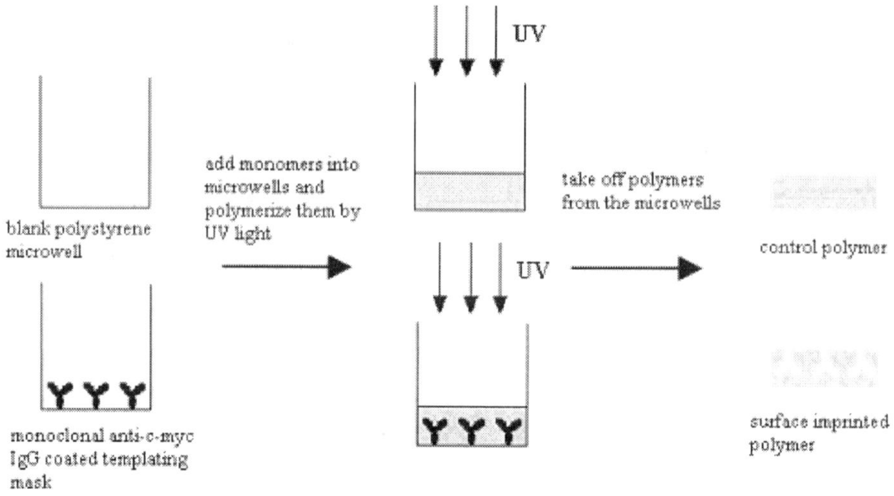

Figure 2: Formation of imprinted and control materials by polymerization in the presence of immobilized target molecule or an unmodified surface.

2.4 Evaluation of recognition ability

After polymerization, the surface imprinted materials and non-imprinted, control materials were peeled away from the microwells. The materials were placed into Corning non-binding microplates with the imprinted surface facing upwards. Next, the polymers were treated with 200 µl Tris buffered saline (pH 7.5) for 1 hour at 4 °C to remove any unreacted monomers and solvent from the polymer networks. After dialysis, the polymers were incubated at 4 °C for 40 minutes in an IgG solution containing enzyme linked conjugates of either the target molecule, structural analogues, or a combination of the two. The enzyme linked IgG solution was removed from the microwells after incubation and the materials were rinsed with PBS-Tween 20 buffer at pH 7.4 for 10 times to remove any unbound proteins. Next, 100 µl of 1-step PNPP was added into each microwell to react with any bound alkaline phosphatase for 20 minutes at room temperature in darkness. The solution was diluted by 100µl fresh DI water after the reaction and 70µl of the diluted solution was taken away to be tested in a microplate reader (Beckman Coulter, DTX 880) at 405 nm. The analysis procedure was done in a manner analogous to ELISA.

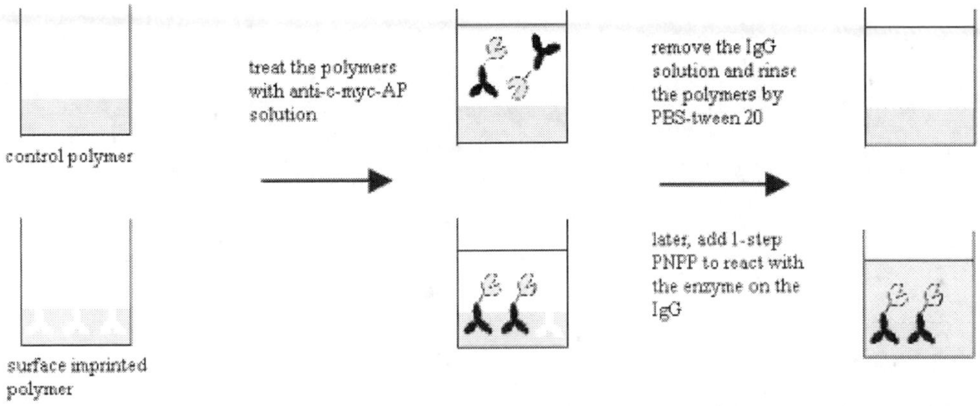

Figure 3: Protein binding was quantified using absorbance measurements of enzyme-linked molecules reacting with substrate

2.5 Material characterization via scanning electron microscope (SEM)

The surface morphology of both the imprinted and non-imprinted materials was visualized with the use of SEM. Specimens were coated in carbon for 30 seconds before using. The chosen scales were 500 and 300 nm and the accelerating voltages for these two scales were 150k and 80k respectively. The working distance was 5.3 mm and accelerating voltage was 5.0 kV.

2.6 Binding efficiency analysis in the addition of a blocking protein

A second protein, rabbit IgG, was introduced during incubation to test whether this species would compete for sites of non-specific adsorption. It was proposed that this addition of a competing, but "silent" (i.e. not enzyme conjugated) molecule would improve the overall binding ratio. In this set of experiments, rabbit IgG was added at various concentrations alongside the target molecule, anti-c-myc-AP.

The polymer preparation procedure was same as 2.3. After dialysis, the polymers were incubated in different antibody solutions as following: Tris buffered saline solution with 0.0016 mg/ml anti-c-myc-AP; Tris buffered saline solution with 0.0016 mg/ml anti-c-myc-AP and 0.0011 mg/ml rabbit IgG; or Tris buffered saline solution with 0.0016 mg/ml anti-c-myc-AP and 0.0022 mg/ml rabbit IgG. The specific binding analysis was identical to 2.4.

3. Results and Discussion

3.1 Material characterization

In Figure 4, SEM images of the surface of both the control and recognitive materials are shown. Both materials gave almost identical images suggesting little physical effect of the imprinting on the surface morphology. As a result, chemical analysis was used in the following to quantitatively determine the recognition ability of the surface imprinted material.

Surface of recognitive materials:

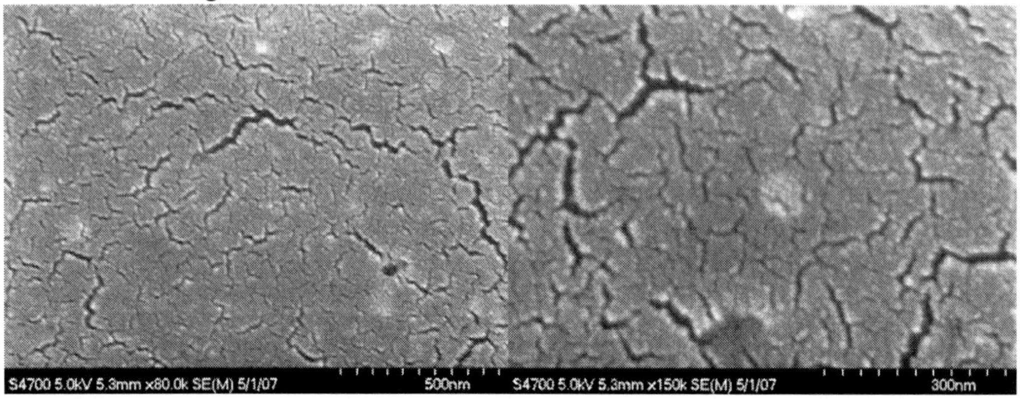

Surface of control materials:

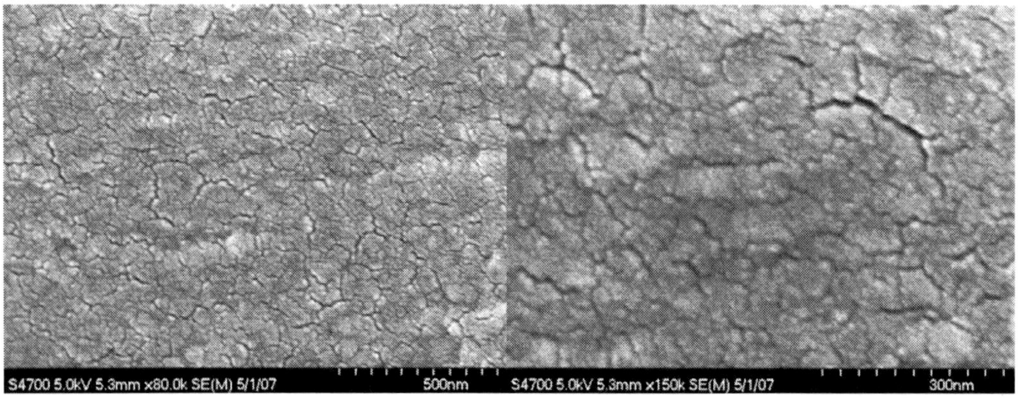

Figure 4: SEM images of recognitive (top) and control (bottom) materials. Scale bars are 500 nm (left images) and 300 nm (right images).

3.3 Protein binding analysis and the effect of a blocking protein

The experimental results are presented in Figure 5. The absorbance ratio between imprinted and non-imprinted materials incubated in a solution with 0.0011 mg/ml rabbit IgG and 0.0016 mg/ml anti-c-myc-AP was 226%. To compare with this, the absorbance difference between imprinted and non-imprinted polymers incubated in 0.0016 mg/ml anti-c-myc-AP solution was 189%. When 0.0022 mg/ml Rabbit IgG was added into 0.0016 mg/ml anti-c-myc-AP solution, the binding ratio between imprinted and non-imprinted polymers was increased to 406%. As a result, the addition of the blocking protein helped the absorbance ratio enhancement between imprinted and non-imprinted polymers. With the increasing of blocking protein in the anti-c-myc-AP solution, the binding ratio between imprinted and non-imprinted polymers became larger and larger.

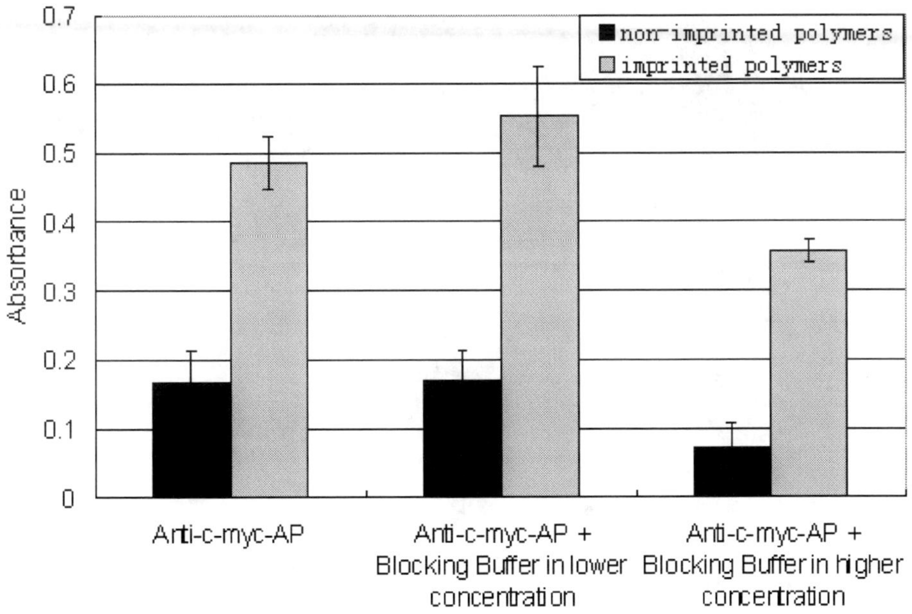

Figure 5: Protein binding study and the effect of a blocking protein

3.2 Binding analysis of MIP prepared with various targets

According to the experimental results shown in Figure 6, there was no significant binding when the materials were incubated in a solution containing only RAP IgG-AP and rabbit IgG. It was clear that no specific recognition of either anti-c-myc-AP or RAP IgG-AP when the polymers synthesized in the 1 % BSA coated microplates. The results also indicated that the surface imprinted polymers made on the template mask could specifically recognize and bind the template molecules out of a solution containing other competitive molecules.

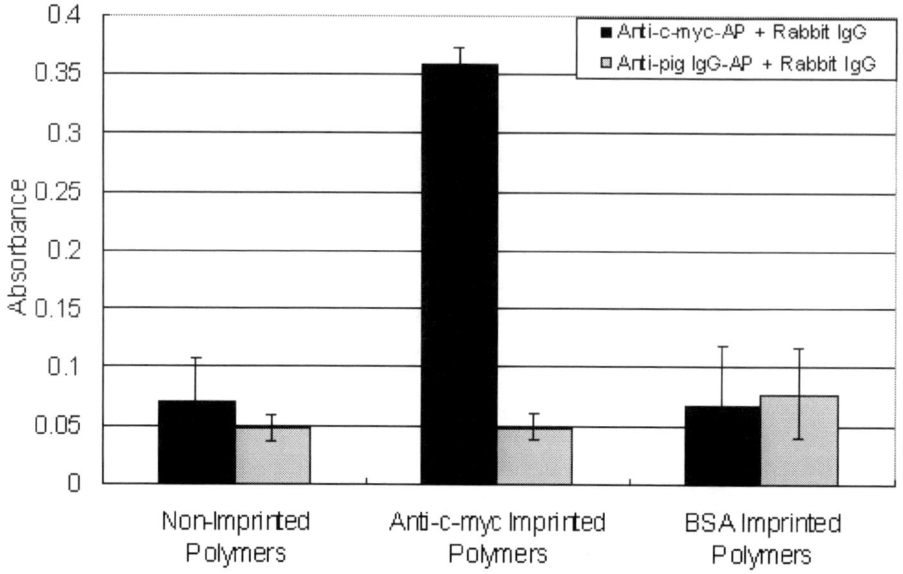

Figure 6: Binding of test immunoglobulins on various imprinted materials.

4. Conclusions

The newly developed surface molecular imprinting technique allows for the production of materials capable of specifically recognizing and binding biological macromolecules. These materials were made in the presence of a monoclonal immunoglobulin which acted as a template. Once produced, the materials demonstrated selectivity toward the template molecule out of a host of competitive species. The addition of a blocking protein enhances the selectivity of the imprinted polymers, as these molecules competed primarily for non-specific adsorption sites. Non-imprinted, control materials revealed no significant binding toward either template molecule or its competitive species. Moreover, materials produced with other template molecules showed no preference for the target molecule. The template mask removed from the imprinted polymers can be reused, therefore the cost of this technique could be further decreased.

5. References

[1] RJ Ansell, O Ramstrom, and K Mosbach. Towards artificial antibodies prepared by molecular imprinting. Clinical Chemistry 42:9, 1506-1512 (1996).

[2] DS Hage. Immunoassays. Anal Chem 67:455R-62R (1995).

[3] RS Yalow. Radioimmunoassay: a probe for the fine structure of biologic systems. Science 200: 1236-1245 (1978).

[4] T Porstmann and ST Kiessig. Enzyme immunoassay techniques: an overview. J Immunol Methods 150:5-21 (1992).

[5] ML Yarmush, AM Weiss, KP Antonsen, DJ Odde, and DM Yarmush. Immunoaffinity purification: basic principles and operational considerations. Biotechnol Adv 1O(3):413-46 (1992).
[6] SE Katz, M Siewierski. Drug residue analysis using immunoaffinity chromatography. J Chromatogr 624:403-9 (1992).
[7] C Yu and K Mosbach. Influence of mobile phase composition and cross-linking density on the enantiomeric recognition properties of molecularly imprinted polymers. J Chromatogr A 888(1-2): 63-72 (2000).
[8] IA Nicholls and JP Rosergren. Molecular imprinting of surfaces. Bioseparation 10:301-305 (2002).
[9] J Svenson and IA Nicholls. On the thermal and chemical stability of molecularly imprinted polymers. Anal Chim Acta 435:19–24 (2001).
[10] NA Peppas and Y Huang. Polymers and gels as molecular recognition agents. Pharmaceutical Research 19: 578-587 (2002).
[11] J-M Lehn. Supramolecular chemistry: Concepts and perspectives. Wiley-VCH. 271pp (1995).
[12] L Ye and K Haupt. Molecularly imprinted polymers as antibody and receptor mimics for assays, sensors and drug discovery. Anal Bioanal Chem 378(8):1887-1897 (2004).
[13] N Kashyap, N Kumar and MNV Ravi Kumar. Hydrogels for pharmaceutical and biomedical applications. Critical Reviews in Therapeutic Drug Carrier Systems 22(2):107-149 (2005).
[14] C Alexander, HS Andersson, LI Andersson, RJ Ansell, N Kirsch, IA Nicholls, J O'Mahony and MJ Whitcombe. Molecular imprinting science and technology: a survey of the literature for the years up to and including 2003. J Mol Recognit 19(2):106-180 (2006).
[15] H Lauten, J Ng and NA Peppas. Selective recognition of angiotensin II for novel drug delivery systems, AICHE2006, US - Japan Joint Topical Conference on Medical Engineering, Drug Delivery Systems and Therapeutic Systems.
[16] T Miyata, M Jige, Y Hishida, K Okawa, Y Ohya, T Ouchi and T Uragami. Preparation of biomolecule-responsive gels by biomolecular imprinting, AICHE 2006, US - Japan Joint Topical Conference on Medical Engineering, Drug Delivery Systems and Therapeutic Systems.
[17] T Miyata, N Asami and T Uragami. A reversibly antigen-responsive hydrogel. Nature 399: 766-769 (1999).
[18] H Shi, WB Tsai, MD Garrison, S Ferrari and BD Ratner. Template-imprinted nanostructured surfaces for protein recognition. Nature 398:593-597 (1999).
[19] A Bossi, SA Pileysky, EV Piletska, PG Righetti and APF Turner. Surface-grafted molecularly imprinted polymers for protein recognition. Anal Chem 73: 5281-5286 (2001)
[20] A Rachkov, N Minoura. Towards molecularly imprinted polymers selective to peptides and proteins. The epitope approach. Biochim Biophys Acta 1544(1-2): 255-266 (2001).
[21] Greg T. Hermanson. Bioconjugate Techniques. Academic Press, 1st edition (1996).

101f Patterning The Surfaces Of Microspheres Using Vapor-Based Polymer Coatings

Joerg Lahann, University of Michigan, Dept. of Chemical Engineering, 3074 H.H. Dow Bldg., Ann Arbor, MI 48109-2136, Hsien-Yeh Chen, Chemical Engineering Department, University of Michigan, 2300 Hayward St, 3434 G.G Brown, Ann Arbor, MI 48109, Erdogan Gulari, Department of Chemical Engineering, University of Michigan, Ann Arbor, MI 48109, and Jean-Marie Rouillard, Chemical Engineering, The University of Michigan Ann Arbor, 2300 Hayward St., 3074 H.H. Dow Building, Ann Arbor, MI 48109.

Compared to the well-equipped arsenal of surface modification methods for flat surfaces, techniques that are applicable to curved, colloidal surfaces are still in their infancies. This technological gap exists, because spin-coating techniques used in traditional photolithographic processes are not applicable to curved surfaces of spherical objects. To generate microstructured colloidal particles we pursued a route that uses (i) Coating of the colloids with a photodefinable polymer, poly[4-benzoyl-p-xylylene- co-p-xylylene], via chemical vapor deposition polymerization; and (ii) spatially controlled surface reaction of the photoreactive coatings using a highly parallel projection lithographic patterning step. The rationale for selecting poly[4-benzoyl-p- xylylene-co-p-xylylene] as base coating was that this CVD-based reactive polymer coating forms homogeneous interlayers on curved surfaces while providing photo-reactive chemical groups for further surface modification (1-2). By replacing spin-coated photoresist with a vapor-deposited, photodefinable polymer coating, we were able to fabricate microstructured colloids with a wide range of surface patterns - including asymmetric and chiral surface structures - that so far were typically reserved for flat substrates. This high-throughput method can yield surface-structured colloidal particles at a rate of about 107 to 108 particles per operator per 1 day. In addition to the controlled fabrication of a series of identical colloids with multiple, biologically distinct binding patches, parallel fabrication of multiple types of distinct colloids is of specific interest to a range of different applications, such as biosensors . High pattern fidelity and excellent contrast are observed suggesting efficient conversion of the photoactive groups in the exposed areas only. In addition, a range of different patterns has been created demonstrating that patterns can be selected with high flexibility. This is in clear contrast to the currently existing methods for patterning of colloids, which are all subject to constrains imposed by the patterning technique itself. Finally, high-resolution patterns can be created on spheres with different radii. Equipped with spatially defined binding pockets, microstructured colloids can engage in programmable interactions, which can lead to directed self-assembly. The ability to create a wide range of colloids with both, simple and complex surface patterns may contribute to the genesis of previously unknown colloidal structures and may have important technological implications in a range of different applications including photonic and phononic materials or chemical sensors.

(1) Suh, K. Y., Langer, R. & Lahann, J. (2004) Advanced Materials 16, 1401-1405. (2) Chen, H.-Y. & Lahann, J. (2005) Analytical Chemistry 77, 6909-6914.

101g 3-D Controlled Synthesis Of Hydrogel Systems Over Micro- And Nanodevices Using Atrp

Hariharasudhan D. Chirra and J. Zach Hilt. Department of Chemical and Materials Engineering, University of Kentucky, 177 F. Paul Anderson Tower, Lexington, KY 40506

Intelligent hydrogel networks have been applied as functional components of various micro- and nanodevices for biomedical and biochemical applications. Conventional methods of the semiconductor industry have been successfully employed for the patterning of such networks in devices with XY control, but the methods for fabricating precise 3-D patterns of hydrogels at the micro- and nanoscale over material surfaces and devices are limited. By amalgamating microcontact printing for XY control and atom transfer radical polymerization (ATRP) for Z (thickness) control, precise patterns of environmentally responsive hydrogels (e.g., N-isopropylacrylamide based systems) have been synthesized over gold and silicon surfaces with 3-D control. The controlled growth and the responsive behavior of the gels to external stimuli were characterized using atomic force microscopy (AFM). Using this method, it is possible to synthesize hydrogels over various surfaces for potential applications in the biomedical field.

121a Use Of Ultra-Low Cost Biodegradable Nanoparticles For The Transfection Of Mammalian Cells

[4] L. K. Limbach, P. Wick, P. Manser, R. N. Grass, A. Bruinink, W. J. Stark, *Environ. Sci. Technol.* **2007**, DOI: 10.1021/es062629t.

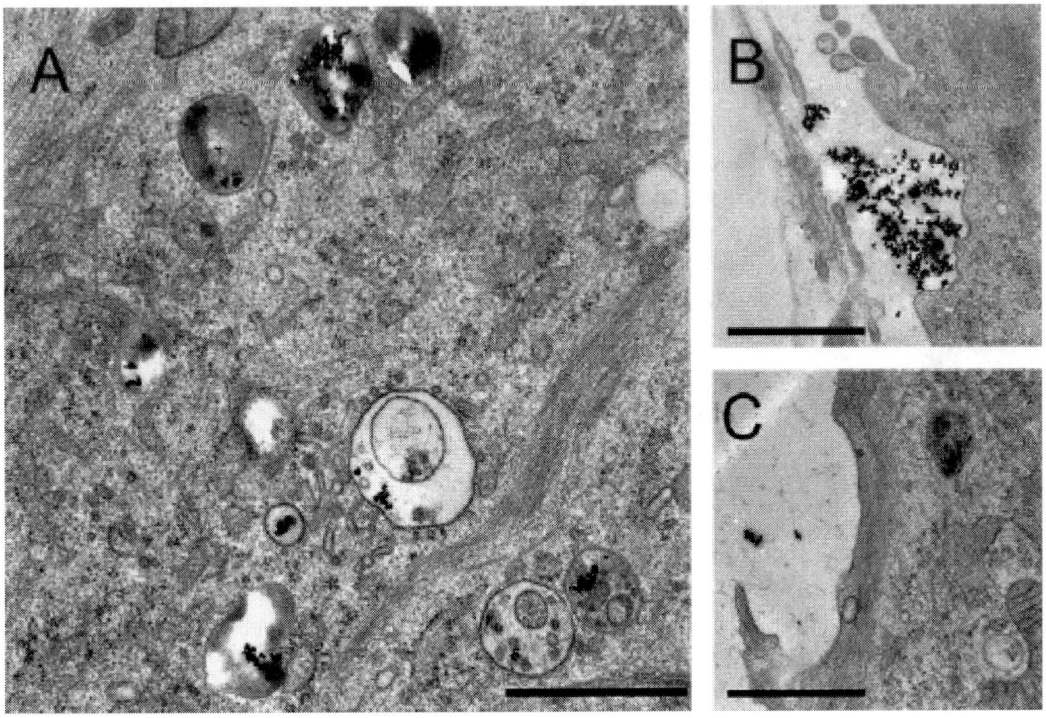

345d Comparison Of Unpromoted Ferrochrome Wgs Catalysts To Those Promoted With Au Or Cu

John S. Coleman and Carl R. F. Lund. Chemical and Biological Engineering, University at Buffalo, SUNY, 303 Furnas Hall, Buffalo, NY 14260

A mechanistic kinetic study of the water-gas shift has been performed using a chromia promoted iron oxide catalyst. The kinetics have been found to be described quite well by either an oxidation-reduction (regenerative, redox) mechanism or by a formate mechanism. To complement the mechanistic kinetics, computational models of the surface intermediates have been generated using density functional theory. The same mechanistic models have also been studied using a copper-promoted ferrochrome catalyst. The addition of copper was found to lessen the severity of the kinetic inhibition by the product carbon dioxide. Computational models where copper was substituted at a lattice iron site were used to discriminate between the mechanistic models. Finally, a gold-promoted ferrochrome catalyst was studied kinetically. At higher temperatures the kinetics of water-gas shift were very similar to the kinetics of an unpromoted catalyst. At lower temperatures, however, it was found to be much more active than the unpromoted catalyst. The low temperature kinetics were studied and a consistent mechanistic model was formulated. The higher temperature results show that it is very difficult to discriminate between the two mechanistic models for water-gas shift on the basis of kinetics alone. The apparent reaction orders were observed to change significantly as the conditions changed. For the low temperature water-gas shift using a gold-promoted catalyst, the present study appears to be one of the first detailed investigations of the kinetics.

Development of Asymmetric Liposomal Nanoparticles for Targeted Delivery of siRNA to Silence Gene Expression

Jeremiah Whittenton [1], Sivaram Harendra [2], Ramanan Pitchumani [3], Tu Nguyen [1], Kishore Mohanty [1], Cumaraswamy Vipulanandan [2] and Sundararajah Thevananther [4]

[1] *Chemical and Biomolecular Engineering, University of Houston, S222 Engineering Bldg. 1, Houston, TX 77204-4004*

[2] *Civil and Environmental Engineering, University of Houston, N107 Engineering Bldg. 1, Houston, TX 77204-4003*

[3] *Biomedical Engineering, University of Houston, N207 Engineering Bldg. 1, Houston, TX 77204-4007*

[4] *Pediatrics, Baylor College of Medicine & Texas Children's Liver Center, One Baylor Plaza, MC3-3391, Houston, TX 77030*

Introduction

It is estimated that hepatocellular carcinoma (HCC) is the fifth most common type of cancer and is the fourth leading cause of cancer-related deaths worldwide. Advances in the molecular understanding of HCC have raised the possibility of silencing cyclin D1 gene expression, which is over-expressed in HCC, by cyclin D1 specific siRNAs. This approach might provide an improved therapy over less effective and toxic systemic therapies. Yet the targeted delivery of siRNA to HCC is challenging since conventional techniques using liposomes and cationic polymers have low encapsulation efficiencies and/or high toxicity. The goal of this study is to develop asymmetric liposomal nanoparticles to overcome these deficiencies.

Liposomes are vesicular colloidal particles composed of self assembled amphiphilic molecules. Amphiphiles are molecules that contain two groups with different solubility. The hydrophilic group, often referred to as the polar head, is "water loving" whereas the tail group is hydrophobic or "water fearing". Amphiphiles can self-assemble to form ordered structures in aqueous solutions. Single chain amphipiles, such as soaps and detergents, can form small spherical structures called micelles in which the polar head groups shield the hydrophobic chains from the bulk aqueous solution. Many natural amphipiles, such as lecithin (diacyl phosphotidylcholine), have two hydrophobic tails which geometrically prohibit them from forming micelles. Instead, these molecules can self-assemble into bilayers with the hydrophilic head groups pointing outward and the hydrophobic chains inward and separated from the bulk aqueous solution.[1] Bilayered lamellae have their edges exposed to water, therefore at lower concentrations they form closed spherical lipid vesicles, called liposomes, to minimize their free energy. Liposomes can be used as a vehicle for drug delivery because they sequester an internal aqueous compartment containing the drug of interest whereas the external aqueous environment in which they freely float does not.[2]

Liposomes have gained considerable interest as drug delivery tool for therapy of cancer and infectious diseases. They are advantageous because they spare healthy tissue from harmful exposure to toxic drugs by delivering the drug to specific sites while increasing systematic circulation time by reducing their removal by the reticuloendothelial system.[3,4] Specific accumulation of liposome particles in tumor cells may be achieved by attachment of targeting agents to the liposome's surface, such as antibodies or RGD peptides.[5,6] Efficient liposome drug delivery systems need to have good colloidal, chemical, and biological stability. Liposomes are non-equilibrium structures and are less sensitive to external changes than equilibrium structures like micelles. Hence colloidally stable liposomes often work well in pharmaceutical applications. The molecular architecture of biological lipid bilayer membranes is highly asymmetric with respect to the composition of membrane proteins, phospholipids, and fatty acids from one leaflet to the other. The asymmetrical distribution of membrane phospholipids across the lipid bilayer seems to be responsible for cell functions like coagulation, membrane fusion, and stability.[7]

Asymmetric liposomes can be produced using an inverse emulsion technique to independently form each leaflet of the bilayer through oil-water interfaces.[8,9] Such an approach should have several advantages. The encapsulated cargo remains distinct from the hosting buffer leading to increased loading efficiencies. It enables fabrication of unilamellar vesicles with a wide range of size through utilizing a single technique.

The objective of this study is to produce liposomes using an inverted emulsion technique and encapsulate fluorescently labeled 21-mer DNA oligo (a cheaper substitute for siRNAs).

Results and Discussion

Figure 1 shows fluorescent images of inverse emulsion particles prepared using POPC, DOTAP, and NBD-PC. POPC, which is a neutral lipid, was placed in the inner leaflet to reduce any electrostatic repulsion that the DOTAP lipids might have with one another. NBD-PC, a fluorescently labeled lipid, was added at 1 mol% to visualize the lipid inner leaflet. The aqueous core contained two fluorescent compounds. The first was Alexa Fluor 350, a water soluble dye for tracking the aqueous solution, and the second was a Cy3 tagged 21-mer DNA oligo for verification of encapsulation. DNA oligo was used instead of siRNA because of its low cost. It is expected that siRNA would perform in a similar manner. This inverse emulsion was extruded with a 5 μm pore diameter membrane. Some particles in the images are larger than 5 μm because some of the inverse emulsion particles, over time, coalesced to from larger particles.

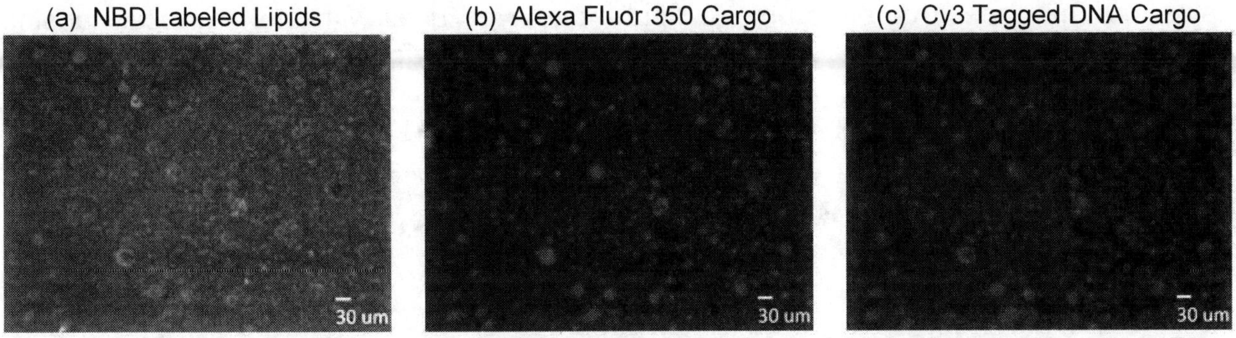

Figure 1. Fluorescent images of the same view field of inverse emulsion containing fluorescent markers. From left to right: (a) Green spheres indicate placement of inner leaflet lipid; (b) Blue spheres indicate the core aqueous solution containing Alexa Fluor 350; (c) Red spheres indicate the placement of a Cy3 tagged 21-mer DNA oligo.

Many attempts were made to form liposome vesicles from inverse emulsions. Liposomes form, yet none have produced evidence showing cargo-encapsulated liposomes in the bottom aqueous phase. For example, when an inverse emulsion sample is extruded at 400 nm, the size distribution of both the emulsion particles and liposomes formed is around 350 nm (measured using DLS) as shown in Figure 2. The fact that we are able to detect the cargo in the inverse emulsion particles, but not in the liposome samples, led us to question the dynamics of the emulsion drops at the interface. To answer this, a visualization cell was designed to view the dynamic interaction between inverse emulsion particles and the oil-water interface.

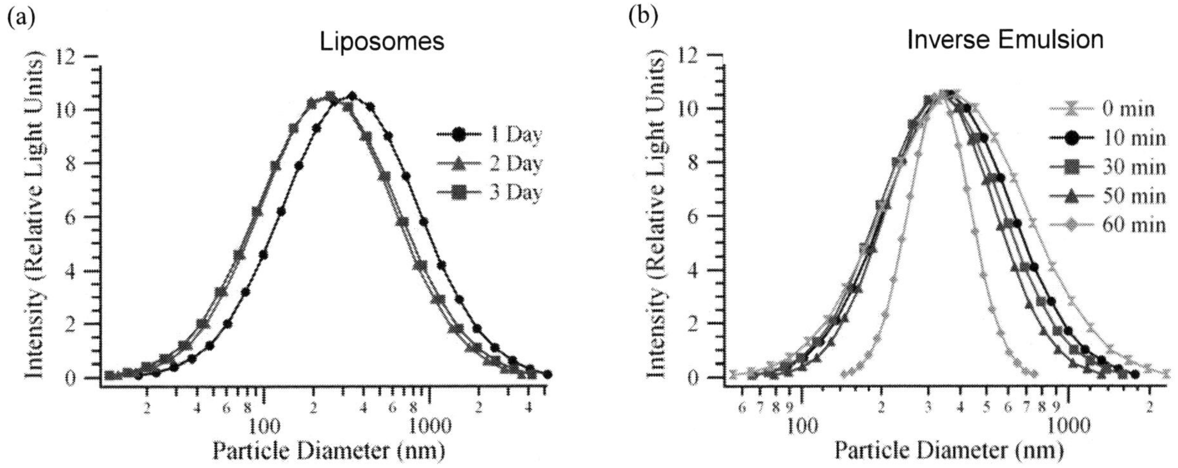

Figure 2. Dynamic light scattering data characterizing the stability as a function of time for liposomes (a) and inverse emulsion (b) prepared using 1 mol% DOTAP and 99 mol% DMPC for the inner leaflet (inverse emulsion) and 100 mol% DMPC for the outer leaflet (Liposomes).

A visualization cell was used to track the fate of large inverse emulsion particles (> 10 μm) as they pass through the intermediate and aqueous phase interface. Figure 3 shows the fate of an inverse emulsion particle (~ 50 μm) as it moves due to gravitational forces through the intermediate phase to the interface where it ruptures. The inverse emulsion was composed of POPC (0.3 mM) and NBD-PC (0.003 mM) in mineral oil with 0.57 mM Alexa Fluor 350 salt as the cargo. The intermediate phase (IP) was identically composed, but without cargo. The first image of the sequence shows the inverse emulsion particle (IEP) just above the interface, which appears as a dark region in the middle of the image. Below the interface is the bottom aqueous phase (AP). Over the course of approximately 40 seconds, the inverse emulsion particle travels approximately 225 µm where it meets the interface. As the particle presses against the interface, it deforms until its lipid monolayer fuses with the interface, causing the particle to rupture instead of passing through the interface to form a bilayer liposome vesicle.

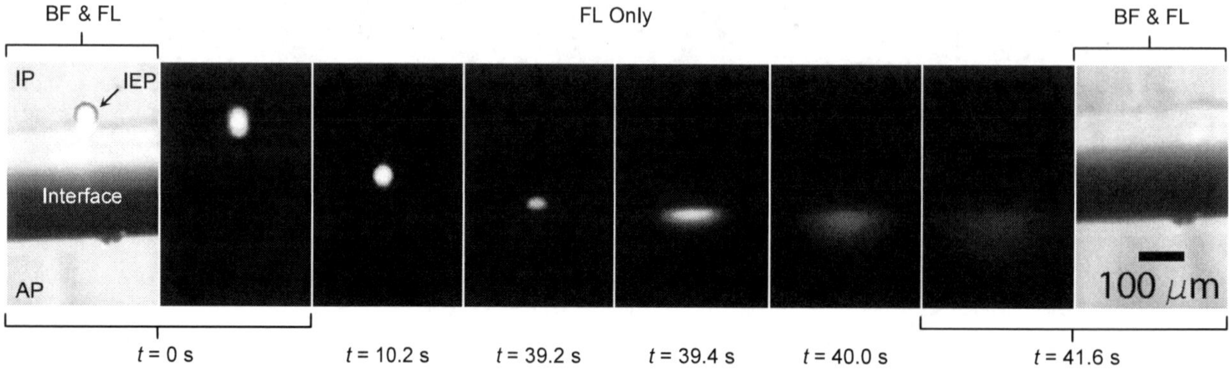

Figure 3. Cross-sectional image time series of an inverse emulsion particle containing Alexa Fluor 350 salt (0.57 mM) as it fuses with the intermediate phase (IP) and aqueous phase (AP) interface to release its contents into the bottom aqueous phase. Inverse emulsion and intermediate phases were made with POPC (0.3 mM) and NBD-PC (0.002 mM) in mineral oil. The first and last images are simultaneous bright field (BF) and fluorescence light (FL) images of the image adjacent to it. The remaining images are solely fluorescence light (FL) images of Alexa Fluor 350 within the inverse emulsion particle.

Inverse emulsion particle fusion with the intermediate and aqueous phase interface was observed under a variety of conditions, including: the use of 0.13-1.0 mM lipid concentration in both the inverse emulsion and intermediate phases; the use of mineral oil, squalene, or dodecane as the oil; the addition of 0.17 mM cholesterol to both oil phases to increase stability; and varying the temperature of the system from 20-50°C. Yet, not all of the observed inverse emulsion particles fused with the interface. Many particles were observed laying stationary at interface and did not pass thorough or fuse with it within the 30 min of observation. Imaging after centrifugation of the visualization cell revealed no formed liposome particles, but did show the disappearance of the stationary inverse emulsion particles.

Furthermore, an experiment was done using a combination of DOTAP (cationic) and DMPC (zwitterionic, but neutral at pH 7.4) lipids in the inverse emulsion and only DMPC in the intermediate phase with Alexa Fluor 350 tagged DNA oligo (0.72 mM) as the cargo. The aim of this experiment was to produce compositionally asymmetric liposomes with cationic lipids in the inner leaflet and neutral lipids in the outer leaflet. It was speculated that the negative charge associated with the phosphate backbone of the DNA oligos would form an electrostatic interaction with the neighboring cationic DOTAP lipids to produce more stable inverse emulsion particles that would stay intact as they passed through the interface. However, imaging with the visualization cell yielded the same inverse emulsion particle fusion results as the POPC systems.

The results repeatedly indicate that the inverse emulsion is not stable or rigid enough to remain intact as it presses against the oil-water interface of the intermediate and aqueous phases.

Conclusions

It was demonstrated that inverse emulsion particles can be produced to encapsulate DNA oligo cargo. It was demonstrated with a visualization cell that large inverse emulsion particles (> 10 μm) are not stable as they pass through the oil-water interface and ultimately rupture before forming liposome particles to decrease yield. Smaller inverse emulsion particles (< 1 μm) however, were detected and characterized using DLS which provided evidence that smaller inverse emulsion particles are more stable as they pass through the oil-water interface. The smaller liposomes were also seen to be relatively stable over a four day period with respect to their size distribution.

Acknowledgments

Work supported by the Alliance for NanoHealth (ANH), Houston, Texas with a grant from the Department of Defense.

References

1. Bangham AD, Horne RW. Negative Staining of Phospholipids and Their Structural Modification by Surface-Active Agents as Observed in the Electron Microscope. *J Mol Biol.* 1964;8:660-8.
2. Papahadjopoulos D, New York Academy of Sciences. *Liposomes and their uses in biology and medicine*. New York Academy of Sciences: New York, 1978; p 462 p.
3. Kirpotin DB, Drummond DC, Shao Y, Shalaby MR, Hong K, Nielsen UB, Marks JD, Benz CC, Park JW. Antibody targeting of long-circulating lipidic nanoparticles does not increase tumor localization but does increase internalization in animal models. *Cancer Res.* 2006;66(13):6732-40.
4. Bohl Kullberg E, Bergstrand N, Carlsson J, Edwards K, Johnsson M, Sjoberg S, Gedda L. Development of EGF-conjugated liposomes for targeted delivery of boronated DNA-binding agents. *Bioconjug Chem.* 2002;13(4):737-43.
5. Ahmad I, Longenecker M, Samuel J, Allen TM. Antibody-targeted delivery of doxorubicin entrapped in sterically stabilized liposomes can eradicate lung cancer in mice. *Cancer Res.* 1993;53(7):1484-8.
6. Temming K, Schiffelers RM, Molema G, Kok RJ. RGD-based strategies for selective delivery of therapeutics and imaging agents to the tumour vasculature. *Drug Resist Updat.* 2005;8(6):381-402.
7. Roy MT, Gallardo M, Estelrich J. Bilayer distribution of phosphatidylserine and phosphatidylethanolamine in lipid vesicles. *Bioconjug Chem.* 1997;8(6):941-5.
8. Yamada A, Yamanaka T, Hamada T, Hase M, Yoshikawa K, Baigl D. Spontaneous transfer of phospholipid-coated oil-in-oil and water-in-oil micro-droplets through an oil/water interface. *Langmuir.* 2006;22(24):9824-8.
9. Pautot S, Frisken BJ, Weitz DA. Production of unilamellar vesicles using an inverted emulsion. *Langmuir.* 2003;19(7):2870-2879.

121d Targeting Polymeric Particles To Sites Of Atherosclerosis

Sudhir P. Deosarkar[1], Ramiro Malgor[2], Jie Fu[3], Leonard D. Kohn[4], Justin Hanes[5], and Douglas J. Goetz[4]. (1) Chemical Engineering, Ohio University, Department of Chemical Engineering, 172 Stocker Center, Athens, OH 45701, (2) Biomedical Sciences, Ohio University, Irvine Hall, Athens, OH 45701, (3) Chemical & Biomolecular Engineering, Johns Hopkins University, 221 Maryland Hall, 3400 N. Charles St, Baltimore, MD 21218, (4) Ohio University, Department of Biomedical Sciences, Athens, OH 45701, (5) The Johns Hopkins University, Department of Chemical & Biomolecular Engineering, 221 Maryland Hall, 3400 N. Charles Street, Baltimore, MD 21218

The increased expression of VCAM-1 at sites of atherogenesis provides a potential target for the selective delivery of drug carriers to sites of atherosclerosis. In this study, we conjugated a VCAM-1 mAb to polystyrene particles and particles made from the biodegradable polymer poly(sebacic acid)-block-polyethylene glycol (PSA-PEG). Protein detection assays (flow cytometry and ELISA) revealed a significant level of VCAM-1 mAb conjugated to the polystyrene and PSA-PEG particles. We investigated the adhesion of the resulting α-VCAM-1 particles to arterial endothelium in a murine model of atherosclerosis (the ApoE$^{-/-}$ mouse). We found that α-VCAM-1 particles exhibit significantly higher adhesion compared to IgG particles. The ratio of α-VCAM-1 particle adhesion to IgG particle adhesion (ligand efficiency), ranged from ~8 to ~32 depending on the polymer used to make up the particles and the artery observed. The ratio of the number of particles adherent in the ApoE$^{-/-}$ mice to the number adherent in the wild type mice (selectivity), was between 1.2 and 1.9 for the polystyrene particles and ~5 for the PSA-PEG particles. Within the ApoE$^{-/-}$ mice, we observed ~2.5 times higher adhesion of the α-VCAM-1 particles to overt plaque regions compared to non-plaque regions of the aorta. These results strongly suggest that drug carriers conjugated with ligands for VCAM-1 will selectively and avidly bind to sites of atherosclerosis via the increased expression of VCAM-1.

Drug Release properties of the enzyme Chondroitinase (cABC) from PLGA nanospheres into the Spinal Cord Injury Area.

Murat Kose, Julie Hasenwinkel and Philip A. Rice
Department of Biomedical and Chemical Engineering
Syracuse, University, Syracuse, NY 13244

ABSTRACT

Spinal cord injury (SCI) has severe impact on millions of people around the world. Finding a cure for the SCI has been difficult but recent research has shown some hope for a cure for patients with SCI.

It is currently known that neurons have ability to regenerate their own injured axons. However, the local environment in the lesion is not hospitable for regeneration due to the inhibitory molecules derived from several cell types in the glial scar. After the Injury, the Chondroitin Sulfate Proteoglycan (CSPG), an extracellular matrix produced by various cells in the area, is the major type of inhibitory molecule in the glial scar.

The axonal regeneration may be enhanced if the amount of CSPG produced after the SCI can be decreased. Since the enzyme chondroitinase (cABC) can degrades CSPG molecules, cABC's controlled release in the region of the spinal cord injury may be of considerable therapeutic value.

Chondroitinase loaded PLGA nanospheres were fabricated using single emulsion method. BSA was not added during the fabrication process, however, small amount of BSA was already exist in the stock enzyme which was purchased from Sigma Aldrich Co. St. Louis MO. Two different total initial cABC concentrations were considered, namely, 300 microliters of 5 unit and 150 microliters of 10 unit cABC. Fabricated nanospheres were freeze dried overnight prior to the release experiments.

A typical release experiment was done by adding 5 mg of cABC loaded nanospheres to 1000 microliters of Phosphate Buffer Solution (PBS) and incubating at 37 ^{o}C. Each set of experiments were done in triplicates. Periodically, 100 microliters of sample from each tube was taken and mixed with 1 ml of 50 μg/ml of Chondroitin Sulfate solution to perform digestion experiments. 100 microliters of fresh PBS were added to the release tubes to compensate the lost volume by removal of the sample. This routine procedure was continued about 30 days.

Digestion experiments were performed by mixing 100 microliters of sample (taken from release experiments) with 1ml of 50 μg/ml of CS solution. At certain time intervals, 100 microliters of sample was taken from digestion tubes and analyzed using DMMB essay. The concentration of cABC released from nanospheres was calculated using the slope of CS digestion profile which was obtained after each sample of release experiment and calibration curve (between slope and cABC concentration) which was prepared earlier. This procedure can be seen in Figure 1.

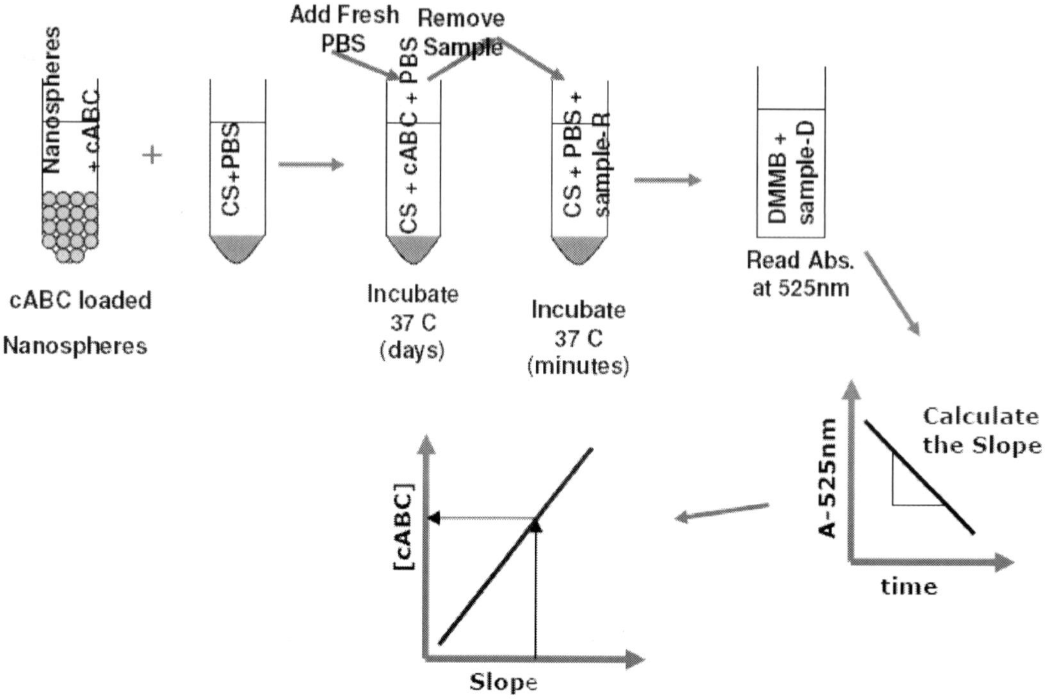

Fig. 1. Calculation of cABC concentration of unknown solution using DMMB Method.

The cumulative cABC release was calculated for both sets of experiments as shown in Figures 2 and 3. The amount of enzyme released from the first set of experiments (300 microliters of 5 unit cABC) is 7.03×10^{-03} units and for the second set of experiments is 7.16×10^{-03} unit. The maximum amount of enzyme that can be possibly released was 6.25×10^{-02} unit as calculated from the total enzyme added during the fabrication process of nanospheres. This indicates that approximately 11% of the original cABC was released from nanospheres and the enzyme was still active. The remaining 89% either remained in the supernatant during the fabrication process or 89% of the incorporated enzyme was not active.

A two step diffusion model has been used to describe the release of cABC from the nanospheres. This model postulates an initial quick (burst) release of enzyme either by diffusion of the enzyme from liquid filled macropores in the nanospheres or desorption of enzyme from the nanosphere surface, followed by a much slower release of enzyme from the nanospheres by diffusion through the PLGA polymer making up the nanospheres. A good fit of the release data to this model was obtained (Figure 4) for burst release of 38% of the total enzyme released, followed by release by diffusion of enzyme through the polymer characterized by an enzyme diffusivity of 3×10^{-17} cm^2/sec.

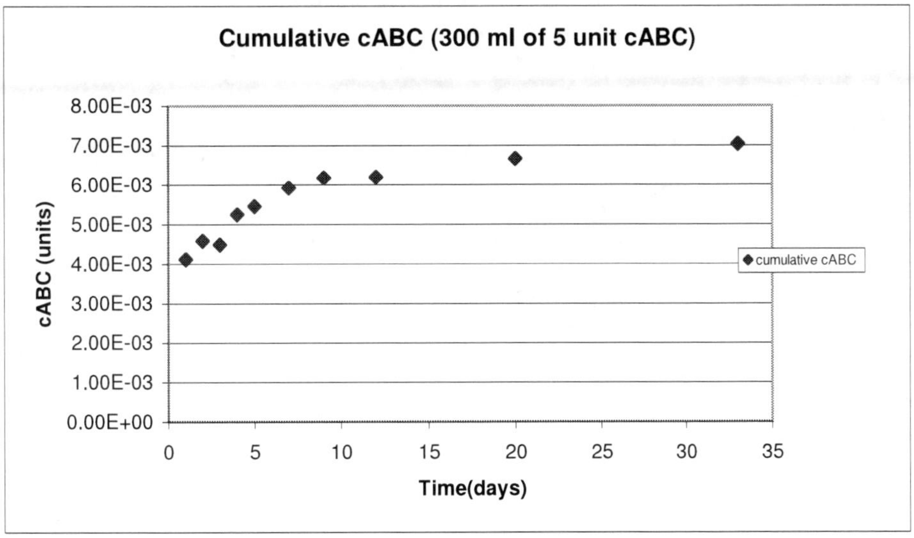

Figure 2. Cumulative cABC release for 300 microliter 5 units of nanospheres

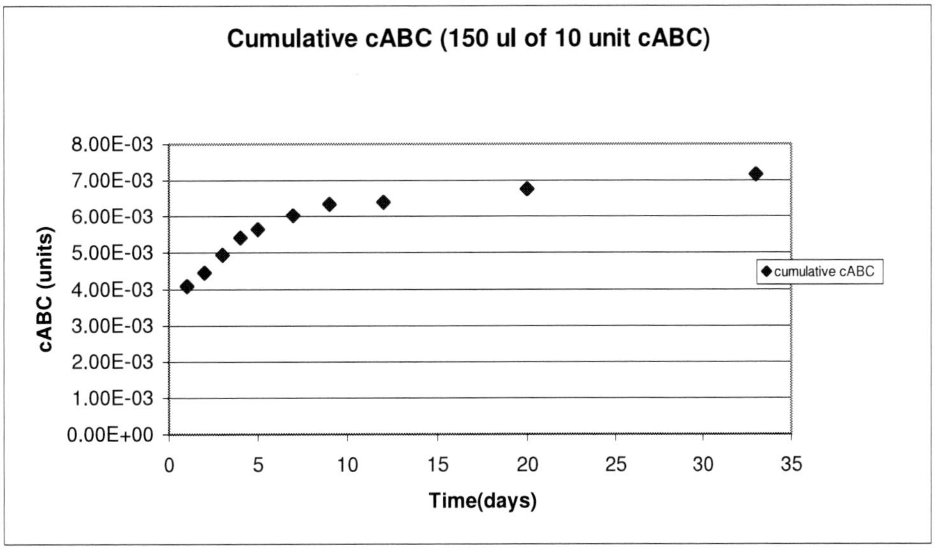

Figure 3. Cumulative cABC release for 150 microliters 10 units of nanospheres

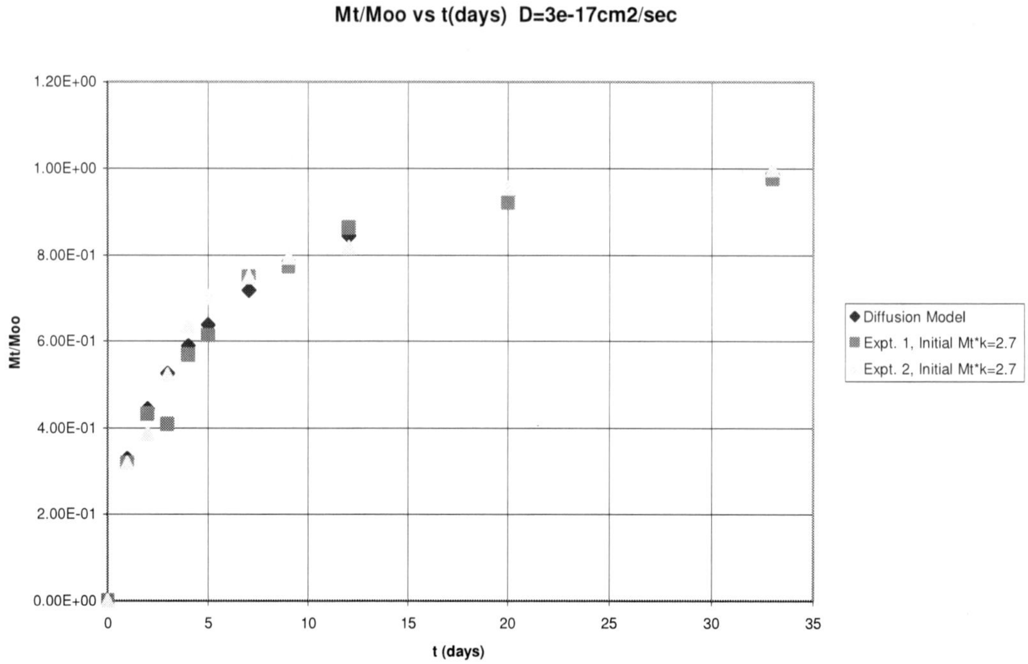

Figure 4. Comparison of the diffusion model and cABC release experiments

121f Engineering Of Peptides For The Targeted Delivery Of Proteins And DNA Into Brain Cells

Shan Gao[1], Melissa J. Simon[2], Barclay Morrison III[2], and Scott Banta[1]. (1) Chemical Engineering, Columbia University, 500 W. 120th street, New York, NY 10027, (2) Biomedical Engineering, Columbia University, 1210 Amsterdam Avenue, New York, NY 10027

One of the major challenges involved in treating brain diseases is the delivery of therapeutic agents into specific types of brain cells. Cell penetrating peptides (CPPs), a group of small and positively charged peptides, have been shown to be able to deliver a wide variety of cargos into mammalian cells in vitro and in vivo while the cell integrity is maintained. TAT, one of most studied CPPs, has been reported to be capable of delivering β-galactosidase into murine brain tissues in vivo. However, the non-specific delivery into other tissues has hampered TAT as a candidate for therapeutic delivery. We are working towards engineering novel specific cell penetrating peptides (SCPPs) to deliver cargos into specific types of brain cells through an innovative directed evolution approach using a plasmid display system to deliver a fluorescent transgene. The transduction ability of TAT for delivery to the brain has been verified by the efficient delivery of green fluorescent protein (GFP) into PC12 cells (31.4±9.2%) and primary astrocytes (10.8±3.4%) through the fusion of TAT at the carboxyl-terminal end of GFP. In addition, HA2, a cell membrane-destabilizing peptide, has been shown to further improve the efficiency of TAT-mediated GFP delivery when it is fused to the amino-terminal end of GFP (64.8±7.8% for PC12 cells and 69.2±15.7% for astrocytes). A plasmid display system used for directed evolution of SCPPs has been established. However, the delivery efficiency of the plasmid display complex requires improvements to enable our directed evolution experiments. The pronounced negative charge of the plasmid DNA likely inhibits the transduction efficiency of TAT. In addition, the size dependence of the delivered DNA cargo on TAT-mediated delivery is being explored using linear DNA fragments with lengths ranging from 20 to 1000 base pairs. The insights gained through these efforts are enabling us to redesign our system for the directed evolution of novel SCPPs.

121g An A-B-C Triblock Copolymer Micelle-Based Approach For Intracellular Delivery Of Gene-Silencing Sirna

Dana J. Gary and You-Yeon Won. School of Chemical Engineering, Purdue University, 408 Stadium Mall Drive, West Lafayette, IN 47907

Following the discovery of the RNA interference (RNAi) mechanism by Fire, Mello and colleagues in the late 1990s [1], there has been explosive interest in the application of short, interfering RNA (siRNA) in the treatment of genetic diseases. These duplex nucleic acids, about 21 base pairs in length, are capable of down regulating harmful genes in mammals with high sequence-specificity but they require a safe and effective delivery vehicle to carry them to the site of action. An ideal carrier would be able to protect the siRNA from enzymatic degradation in the blood stream, be selectively internalized by target cells and then release the siRNA only once it has reached the cytosol, where it can initiate the RNAi pathway. Polymers, particularly polycations, are very popular non-viral gene carriers because they can spontaneously associate with the negatively-charged phosphate backbone of nucleic acids to form "polyplexes". However, among the drawbacks of these systems is that under physiological salt conditions, where repulsive forces between like-charged particles are screened, aggregation of cationic polyplexes can result [2]. Conjugation of poly(ethylene glycol), or PEG, to cationic polymers has been shown to confer several advantageous properties including colloidal stability and reduced aggregation of complexes due to steric shielding [2-4] and increased hydrophilicity/water solubility [4]. These advantages come at price, however, as "PEGylation" is also known to weaken the strength of binding between polycations and plasmid DNA [3-9], leaving the DNA susceptible to enzymatic degradation [4]. Fortunately, polymeric vehicles designed for delivery of DNA for gene therapy have been widely studied over the years, and knowledge from this field has been instrumental in identifying issues in siRNA delivery. In a recent study, we sought to evaluate the potential of three polycation-based gene carriers by studying their relative strengths of binding to siRNA, enzymatic protection and cytotoxic effects. This was illustrated with the use of two conventional gene-delivery approaches, (1) the polycationic homopolymer, poly(2-(dimethylamino)ethyl methacrylate) or PDMAEMA, and (2) a PEGylated polycation, PEG-PDMAEMA. Also, we introduce (3) a micelle derived from an A-B-C triblock copolymer, PEG-PnBA-PDMAEMA (PnBA denotes poly(n-butyl acrylate)). The micelle chemistry we present here has not been previously reported in the literature and differs from other micelle-based approaches because the polycation is indirectly connected to the PEG block through a hydrophobic segment. Contrary to what has been shown for DNA, our studies indicate that in the case of siRNA, PEGylation of polycations does not weaken electrostatic interactions nor promote enzymatic degradation when compared to their un-PEGylated counterparts. We hypothesize that because of molecular weight and/or chemistry, siRNA associates with polymers in a way such that the formed complexes are not adversely affected by PEGylation. Although the triblock micelles have demonstrated a performance comparable to the other formulations with respect to cytotoxicity and in their ability to bind and protect siRNA from enzymatic degradation, we expect that they will outperform the other formulations in the areas of in-vivo pharmacokinetics and cellular transfection. Because homopolymer and diblock complexes with siRNA are significantly smaller than micelle/siRNA complexes (<10 nm vs.

~70 nm), they have a higher probability of being eliminated from the bloodstream before eliciting their biological effects. Thus, micelle-based vehicles appear to be more promising candidates for efficient systemic delivery of gene silencing siRNA when compared with conventional polymer-based approaches. If time allows, we will also discuss our results from in vitro and in vivo gene silencing experiments. References 1. Fire, A., et al., Potent and specific genetic interference by double-stranded RNA in Caenorhabditis elegans. Nature, 1998. 391(6669): p. 806-811. 2. Mishra, S., P. Webster, and M. E. Davis, PEGylation significantly affects cellular uptake and intracellular trafficking of non-viral gene delivery particles. European Journal of Cell Biology, 2004. 83(3): p. 97-111. 3. Glodde, M., S.R. Sirsi, and G.J. Lutz, Physiochemical properties of low and high molecular weight poly(ethylene glycol)-grafted poly(ethylene imine) copolymers and their complexes with oligonucleotides. Biomacromolecules, 2006. 7(1): p. 347-356. 4. Mullen, P.M., et al., Strength of conjugate binding to plasmid DNA affects degradation rate and expression level in vivo. Biochimica Et Biophysica Acta-General Subjects, 2000. 1523(1): p. 103-110. 5. Pun, S.H. and M.E. Davis, Development of a nonviral gene delivery vehicle for systemic application. Bioconjugate Chemistry, 2002. 13(3): p. 630-639. 6. Nguyen, H.K., et al., Evaluation of polyether-polyethyleneimine graft copolymers as gene transfer agents. Gene Therapy, 2000. 7(2): p. 126-138. 7. Gary, D.J., N. Puri, and Y.Y. Won, Polymer-Based siRNA Delivery: Perspectives on the Fundamental and Phenomenological Distinctions from Polymer-Based DNA Delivery. accepted in Journal of Controlled Release, 2007. 8. Garrett, S.W., et al., Synthesis and characterisation of polyamine-poly(ethylene glycol) constructs for DNA binding and gene delivery. Bioorganic & Medicinal Chemistry, 2000. 8(7): p. 1779-1797. 9. Oupicky, D., R.C. Carlisle, and L.W. Seymour, Triggered intracellular activation of disulfide crosslinked polyelectrolyte gene delivery complexes with extended systemic circulation in vivo. Gene Therapy, 2001. 8(9): p. 713-724.

121h A Novel Selenium-Carbon Composite Nanomaterial For Cell Delivery Of Selenium

Love Sarin[1], Aihui Yan[1], Vanesa Sanchez[1], Agnes Kane[2], and Robert H. Hurt[3]. (1) Brown University, Providence, RI 02912, (2) Department of Pathology and Laboratory Medicine, Brown University, 70 Ship street, Providence, RI 02912, (3) Engineering Division, Brown University, 182 Hope street, Providence, RI 02912

Elemental selenium is a promising therapeutic agent in the treatment of a variety of diseases. Here we demonstrate the synthesis of a novel nanomaterial which is able to deliver selenium to cells. Carbon nanoparticles are first synthesized by a dilute ultrasonic spray pyrolysis method, and subsequently they are surface doped with elemental selenium nanoparticles by an aqueous phase selenite salt reduction with glutathione. As an alternative approach, pure selenium nanoparticles are synthesized and stabilized in the presence of bovine serum albumin. The two competing selenium-containing nanoforms are characterized by electron microscopy, dynamic light scattering, elemental analysis, and X-ray photoelectron spectroscopy. Biological tests are underway to compare the effects and mechanism of cell delivery of these forms of nanostructured selenium.

134a In Vivo Hard Tissue Response and Degradation of Porous Fumarate-Based Polymer/alumoxane Nanocomposites for Bone Tissue Engineering

Amit S. Mistry[1], Quynh Pham[1], Corinne Schouten[2], Tiffany Yeh[1], Antonios G. Mikos[1], and John A. Jansen[2]. (1) Bioengineering, Rice University, P.O. Box 1892, MS 142, Houston, TX 77251-1892, (2) Periodontology and Biomaterials, Radboud University Nijmegen Medical Center, Nijmegen, 6500 HB

Introduction. Scaffolds for bone tissue engineering are designed with consideration of mechanical properties, osteoconductivity, degradability, and biocompatibility. In this work, porous scaffolds composed of fumarate-based polymers and alumoxane nanoparticles were fabricated and then evaluated for in vivo hard tissue response and in vivo degradation.

Methods. Scaffolds were prepared by photo-crosslinking and salt-leaching from three materials: poly (propylene fumarate)/propylene fumarate-diacrylate (PPF/PF-DA) polymer alone; PPF/PF-DA with a 1 wt. % loading of boehmite microparticles (macrocomposite group); and PPF/PF-DA with a 1 wt. % loading of surface-modified alumoxane nanoparticles (nanocomposite group). The boehmite particles in the macrocomposite are known to aggregate into large clusters that weaken a material. On the other hand, the surface-modified alumoxane nanoparticles in the nanocomposite significantly improved the mechanical properties of the PPF/PF-DA polymer in a previous study [1]. Additionally, a low molecular weight (MW) PPF polymer group was tested as a control for degradability. Scaffolds were implanted in the lateral femoral condyles of adult goats for 12 weeks and evaluated by micro-computed tomography (micro-CT) and histological analysis.

Results and Discussion. In general, most scaffolds were filled with soft-tissue, some inflammatory elements, and some immature bone tissue after 12 weeks in vivo. Both micro-CT and histology revealed that the presence of alumoxane nanoparticles in PPF/PF-DA had no effect on bone formation in the defects. Some degree of bone growth into the porous network of scaffolds was observed in 60-70% of sections from the PPF/PF-DA polymer, macrocomposite, and nanocomposite groups. Bone growth was also observed in 23% of sections from the degradation control group. The response at the interface between the implant and the surrounding tissue was typically characterized by a thin fibrous capsule. Direct contact between the exterior of the implants and the surrounding bone tissue was observed in 30-40% of sections from PPF/PF-DA-based materials and in 58% of sections from the control group. Once again, the presence of alumoxane nanoparticles had no effect on the bone tissue response at the implant-tissue interface. In vivo degradation was scored based on observations of fragmentation and polymer dissolution at the edges of scaffolds. Though very limited degradation was observed in PPF/PF-DA-based scaffolds, the presence of alumoxane nanoparticles in the polymer was found to have no detrimental influence on in vivo degradation. In summary, the incorporation of surface-modified alumoxane nanoparticles into fumarate-based polymers provides mechanical reinforcement without any detrimental effect on hard tissue response or in vivo degradation.

Reference. 1. Horch, RA, Shahid, N, Mistry, AS, Timmer, MD, Mikos, AG and Barron, AR. Nanoreinforcement of poly(propylene fumarate)-based networks with surface modified alumoxane nanoparticles for bone tissue engineering. Biomacromolecules 5, 1990-1998 (2004).

Web Page: www.ruf.rice.edu/~mikosgrp/

Nanostructured Titania/PLGA Composite Scaffolds Improve Cytocompatibility and Mechanical Strength for Better Bone Regeneration

Huinan Liu and Thomas J. Webster
Division of Engineering, Brown University, 182 Hope Street, Providence, RI 02912, USA

ABSTRACT

The design of synthetic scaffolds is crucial for replacing functions to diseased or injured bones. The desired scaffolds should be able to rapidly induce new bone formation, should possess initial sufficient mechanical strength for load-bearing applications before new bone has been regenerated, and should maintain their mechanical integrity until new bone formation has occurred. In this study, a bio-inspired ceramic/polymer nanocomposite was developed to serve as a mirror template for new bone growth. Specifically, nano-sized titania particles were dispersed in a model polymer (PLGA or poly-lactide-co-glycolide) matrix using controlled sonication. Results of this study showed that the nano-titania/PLGA composites provided better biological and mechanical properties. For example, in vitro studies provided evidence of increased osteoblast (bone-forming cell) adhesion and calcium-containing mineral deposition on well-dispersed nano-titania in PLGA composites compared to pure PLGA scaffolds and the more agglomerated nano-titania in PLGA scaffolds. Moreover, well-dispersed nano-titania in PLGA increased the compressive modulus of scaffolds compared to pure PLGA scaffolds and the more agglomerated nano-titania in PLGA scaffolds. The addition of titania nanoparticles in PLGA also decreased the weight loss of scaffolds, reduced harmful acidic pH changes during PLGA degradation, and prolonged mechanical integrity of such scaffolds. In summary, the combination of PLGA with a strong and biocompatible well-dispersed nano-titania phase created better materials for bone regeneration.

Keywords: nanocomposite, scaffold, ceramic, polymer, titania, PLGA, surface roughness, osteoblast, degradation, mechanical property, orthopedics, bone regeneration.

INTRODUCTION

Natural bone is a nanostructured composite material composed of a ceramic phase (carbonated hydroxyapatite) and an organic phase (mainly collagen) [1]. Ceramic/polymer composites have been considered as the third-generation orthopedic biomaterials due to their closer-matched properties with natural bone compared to first (metals or metal alloys) and second generation (ceramics) bone substitute materials [2].

In the past decade, much attention has been paid to the nanophase ceramic/polymer composites after Webster et al. first reported that osteoblast (bone-forming cell) adhesion significantly increased on nanophase ceramics (such as alumina and titania) compared to the respective conventional ceramics [3]. Webster et al further confirmed that osteoblast long-term functions (such as proliferation and differentiation) were also consistently enhanced on nanophase ceramics (alumina, titania and hydroxyapatite) compared to respective conventional ceramics [4].

One of the key factors that influence properties of composite materials is dispersion status of ceramics in polymer composites. From a biomimetic perspective, considering bone at the nanostructural level, hydroxyapatite crystals located specifically in the discrete spaces within the collagen fibrils and this unique architectural arrangement grants natural bone distinctive biological and mechanical properties [2]. Because agglomeration significantly increases as ceramic particles decrease into the nanometer regime and as the percentage of ceramic particles increases to more than 2 wt%, the objective of the present in vitro study was to disperse nano-titania in PLGA and compare biological and mechanical properties of well-dispersed to agglomerated nano-titania/PLGA composites and PLGA controls.

MATERIALS AND METHODS

Preparation of nanocomposites: PLGA pellets (50/50 wt.% poly(dl-lactide/glycolide); Polysciences) were dissolved in an organic solvent and nanophase titania powder (Nanophase Technologies) was then dispersed in the PLGA solution to give a 70/30 polymer/ceramic weight ratio. After sonication, the suspension was cast into a mold, evaporated in air at room temperature for 24 hours and dried in an air vacuum chamber at room temperature for 48 hours. The dispersion status of final composite scaffolds was controlled by sonication settings. These PLGA/titania composites (PTC) were termed as PTCa (a=agglomerated) and PTCd (d=dispersed) according to their dispersion states.

Preparation of control materials: Pure PLGA and nanophase titania compacts were used as controls. PLGA was prepared by a solvent-casting technique. Nanophase titania compacts were prepared by dry pressing in a tool-steel die via a uniaxial pressing cycle from 0.6 to 3 GPa over a 10 minute period. These titania compacts were then sintered at 600 °C for 2 hours. These compacts were termed as sintered titania compacts (TCS).

Surface characterization: Surface topographies of the nanophase titania/PLGA composites were characterized using a JEOL JSM-840 Scanning Electron Microscope at a 5 kV accelerating voltage. Substrates were sputter-coated with a thin layer of gold - palladium, using a Hummer I Sputter Coater (Technics) in a 100 mTorr vacuum argon environment for 3 min with 0.01 Amp of current. Atomic force microscopy (AFM) was used to characterize the three-dimensional surface features and surface roughness of the materials of interest to the present study. The surface information was collected according to established tapping mode techniques using a Multimode™ SPM (Digital Instruments). All surface roughness data were collected from five different $5 \times 5\ \mu m^2$ AFM scans. The average root mean square value was calculated.

Osteoblast adhesion: Human osteoblasts (ATCC) were seeded at a concentration of 2500 cells/cm^2 onto the scaffolds of interest in Dulbecco's modified Eagle's medium (DMEM; GIBCO) supplemented with 10% fetal bovine serum (FBS; Hyclone) and 1% penicillin/streptomycin (P/S; Hyclone) and then incubated under standard cell culture conditions for 4 hours. After that time period, non-adherent cells were removed by rinsing with PBS and adherent cells were then stained with DAPI nucleic acid stain (Invitrogen); the cell nuclei were thus visualized and counted under a fluorescent microscope.

Osteoblast long-term functions: Osteoblasts were seeded at a density of 100,000 cells/cm^2 onto the scaffolds of interest and were cultured in DMEM supplemented with 10% FBS, 1% P/S, 50 µg/mL L-ascorbic acid (Sigma) and 10 mM β-glycerophosphate (Sigma), under standard cell culture conditions for 7, 14, and 21 days. Cell culture media was changed every other day. At the end of the prescribed time periods, osteoblasts were lysed using three freeze–thaw cycles according to standard protocols to determine total collagen content and calcium deposition.

Degradation of scaffolds: Initial dry scaffolds of interest were weighed (W_0) and sterilized. Then, all the scaffolds were immersed into PBS and were incubated under standard cell culture conditions. The pH of the supernatant buffer was monitored three times a week during the experiments. After 21, 28, and 35 days, specimens were removed from PBS, abundantly rinsed with deionized water to remove the soluble inorganic salt, and dried in an air vacuum chamber at room temperature for 48 hours to reach constant mass. At each time point, samples were weighed (W_t) and the percentage of weight loss (%WL) with respect to incubation time was calculated according to the following equation.

$$\%WL = \frac{(W_0 - W_t)}{W_0} \times 100\%$$

Mechanical test: All composites and PLGA were subject to compressive tests using an Instron 4505 mechanical testing system. Load and displacement data were collected and the compressive moduli were calculated from the stress-strain curves.

Statistical Analysis: All in vitro experiments (such as osteoblast adhesion, long-term functions and degradation of scaffolds) were run in triplicate and repeated three times. Mechanical tests were repeated three times for each type of specimens. Numerical data were analyzed using standard analysis of variance (ANOVA) techniques and Student *t* test; statistical significance was considered at $p<0.05$.

RESULTS

Surface characterization: Scanning electron micrographs suggest that the distribution of nano-titania particles was very different on the surface of the composite scaffolds depending on the sonication setting utilized, as shown in Figure 1. Specifically, there were less titania particles on the surface of PTCa because the agglomerates larger than 100 nm descended faster than the solvent evaporation rate according to the established Stoke's Equation. The amount of surface area occupied by titania increased on the surface of PTCd (10.1%) compared to PTCa (5.7%) because the solvent evaporation was much faster than the sedimentation of well-dispersed titania particles less than 100 nm. To note, the weight ratio of nano-titania to PLGA was the same for PTCa and PTCd. In Figure 1, it is apparent that the surface of PLGA was rather smooth while the surface of TCS was more rough and, thus, more similar to the surface structure of natural bone.

Atomic force microscopy results demonstrated that all the titania/PLGA composites fabricated at different sonication settings had nanometer scale surface roughness from 20 nm to 40 nm according to 5 μm × 5 μm AFM scans. AFM images (in ref [1], not shown here) confirmed SEM results that the dispersion of nano-titania was significantly improved on the surface of the composites when higher sonication power was utilized. Moreover, the surface roughness of all the scaffolds compared to natural bone was plotted in Figure 2. Results showed that: (i) the surface roughness of PTCd was significantly greater than PLGA and PTCa, and (ii) the surface roughness of TCS was the greatest and closest to natural bone.

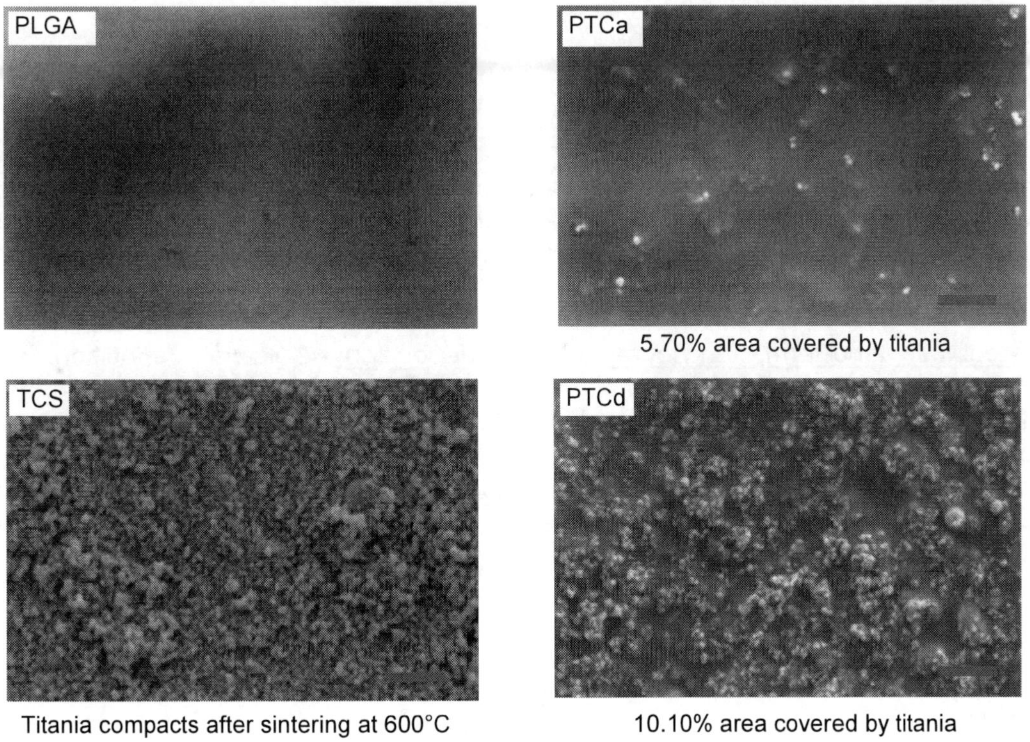

Figure 1: SEM micrographs of nanophase titania/PLGA composite scaffolds: PTCa (a=agglomerated) and PTCd (d=dispersed); and control scaffolds: PLGA and TCS. Magnification bars: 1 μm.

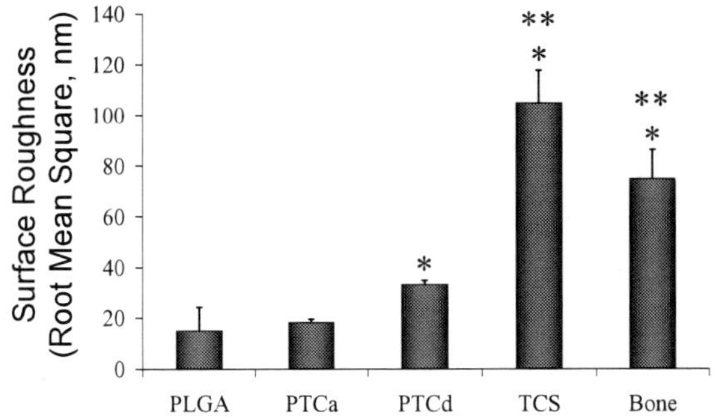

Figure 2: Surface roughness (root-mean-square) of the materials of interest and natural bone. Values are mean ± SEM; n=5; *$p < 0.05$ compared to PLGA and PTCa; and **$p < 0.05$ compared to PLGA, PTCa and PTCd.

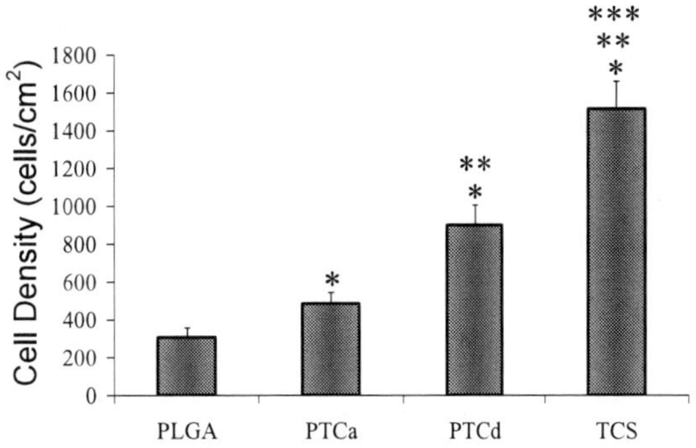

Figure 3: Osteoblast adhesion on the materials of interest. Values are mean ± SEM; n = 3; *$p < 0.05$ compared to PLGA; **$p < 0.05$ compared to PTCa; and ***$p < 0.05$ compared to PLGA, PTCa and PTCd.

Osteoblast adhesion: Adhesion is a critical initial step for long-term osteoblast-material interactions. Results showed that osteoblast adhesion was the greatest on the TCS and was significantly greater on all the composites than on the PLGA, as shown in Figure 3. Moreover, osteoblast adhesion was significantly greater on the PTCd than on the PTCa. Most interestingly, osteoblast adhesion was closely related to surface roughness of scaffolds, if comparing Figure 2 with Figure 3.

Osteoblast long-term functions---collagen synthesis: Total collagen synthesis by osteoblasts is a key indicator of osteoblast proliferation and extracellular matrix synthesis. As shown in Figure 4, promisingly, collagen synthesis increased with longer time periods of culture. For example, total collagen content increased significantly on all the scaffolds after 21 days of culture compared to 7 days of culture. After 21 days of culture, total collagen synthesis was the greatest on the TCS and was significantly greater on the PTCd than on the PLGA and PTCa.

Osteoblast long-term functions---calcium deposition: Calcium deposition is an essential indicator of osteoblast differentiation and mineralization of extracellular matrix. The amount of calcium deposited by osteoblasts increased on all the composites and the TCS after 21 days of culture compared to 7 days of culture. However, calcium deposition on the PLGA did not increase after 21 days of culture compared to 7 days of culture. After 21 days of culture, the amount of calcium deposited by osteoblasts was the greatest on the TCS and was significantly greater on all the composites than on the PLGA.

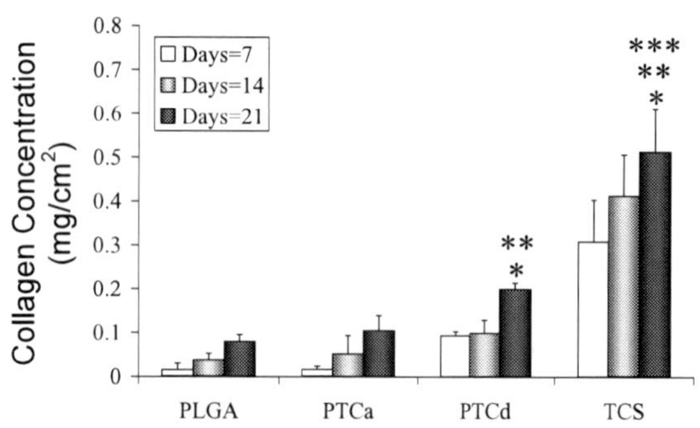

Figure 4: Total collagen content in osteoblasts cultured on the materials of interest. Values are mean ± SEM; n = 3; *$p < 0.05$ compared to PLGA; **$p < 0.05$ compared to PTCa; and ***$p < 0.05$ compared to all the composites and PLGA.

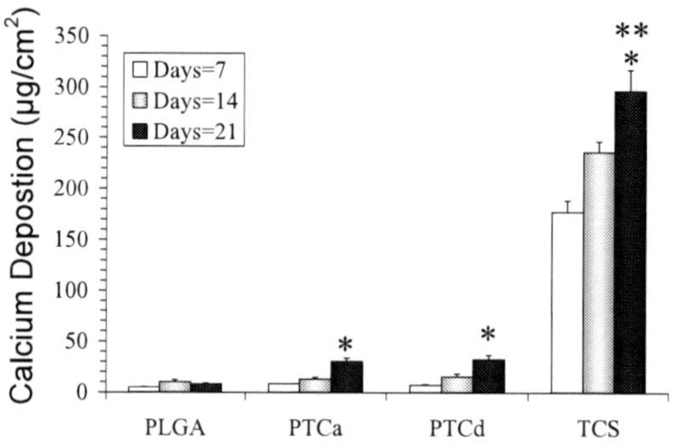

Figure 5: Calcium deposited by osteoblasts cultured on the materials of interest. Values are mean ± SEM; n = 3; *$p < 0.05$ compared to PLGA; and **$p < 0.05$ compared to all the composites and PLGA.

Mechanical properties: Besides all these biological advantages of well-dispersed nano-titania in PLGA composites, these nanocomposites enhanced mechanical properties of scaffolds for potential loading-bearing applications. As shown in Figure 6, the compressive modulus of the PTCd was about 2 times higher than the PTCa and the compressive modulus of the PTCa was about 2 times higher than the PLGA. In comparison to natural bone, the compressive modulus of the PTCd was closer to low-density cancellous bone (~ 10 MPa) than PTCa and PLGA, although it was still lower than high-density compact bone (~ 10 GPa) [5].

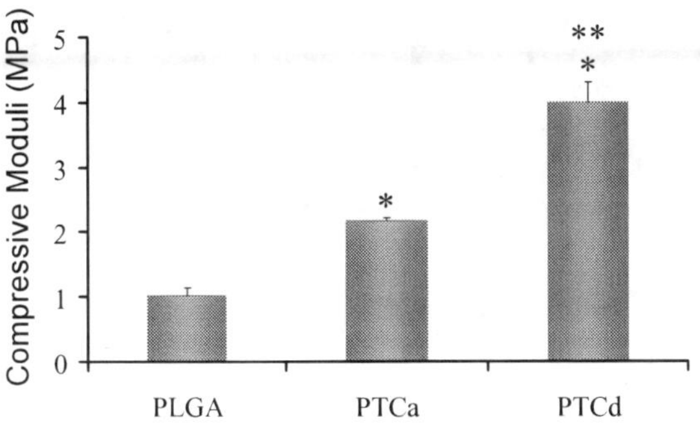

Figure 6: The compressive moduli of the materials of interest. Values are mean ± SEM; n = 3; *p < 0.05 compared to PLGA; and **p < 0.05 compared to PTCa.

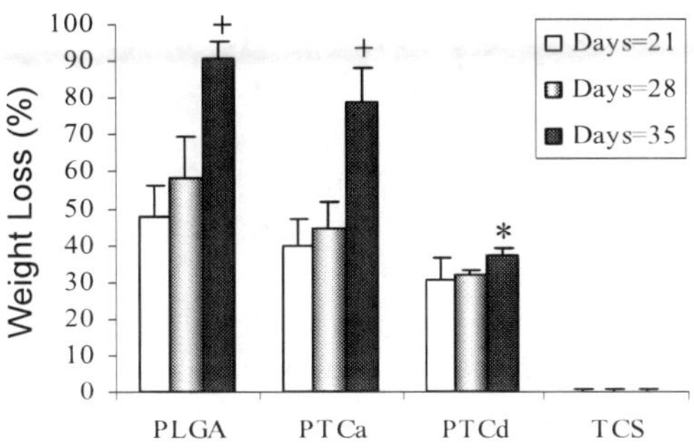

Figure 7: The weight loss of scaffolds incubated in PBS at 37 °C. Values are mean ± SEM; n = 3; *p < 0.05 compared to PLGA and PTCa; and +p < 0.05 compared to respective shorter days of culture.

Degradation of scaffolds: The degradation kinetics of PLGA is another important factor that influences cell responses to the scaffolds. PLGA degrades with random chain scission by ester hydrolysis in a process auto-catalyzed by the generation of carboxylic acid end groups, and titania particles used as inclusions in PLGA can tailor the degradation kinetics of the polymer matrix and prevent acceleration of polymer degradation. As shown in Figure 7, the percentage of weight loss of the PLGA was the greatest among all the scaffolds incubated at respective days. No weight loss was observed on the TCS. Importantly, the weight loss of the PTCd was about 50% less than the PLGA and the PTCa after 35 days of incubation. This indicated that the dispersion status of nano-titania in PLGA composites did play an important role in the degradation behavior of PLGA. Moreover, the buffering effect of nano-titania particles in weight loss was more significant after longer time periods of incubation, which was closely correlated to the pH buffering effect of these nano-titania particles in PLGA. Specifically, the pH drop was 70% for the PLGA while it was only 43% for the PTCd after 35 days of incubation.

CONCLUSIONS

The results of this study demonstrated that nanophase titania has exceptional cytocompatibility with osteoblasts. Specifically, osteoblast adhesion and long-term functions on nanophase titania compacts were the greatest among all the scaffolds. However, titania itself is inherently brittle due to its ceramic nature and is very difficult to fabricate into complex-shaped bone substitutes with acceptable mechanical properties. Therefore, when considering all these results together, the well-dispersed nano-titania/PLGA composite (PTCd) is the best candidate as an orthopedic material for more effective bone regeneration. Moreover, the dispersion of titania nanoparticles in PLGA promoted mechanical properties of scaffolds and also provided better degradation kinetics which favors bone cell survival and physiological functions by decreasing the weight loss of scaffolds, reducing harmful acidic pH changes during PLGA degradation and prolonging the mechanical integrity of such scaffolds. In conclusion, the combination of PLGA with a strong and biocompatible well-dispersed nano-titania phase can be very promising scaffold materials for more effective orthopedic tissue engineering applications.

ACKNOWLEDGEMENTS

The authors would like to thank the NSF for a Nanoscale Exploration Research (NER) grant and Christopher Bull for the assistance with Instron 4505 mechanical testing system.

REFERENCES

1 Liu H, Slamovich EB, Webster TJ. Increased osteoblast functions among nanophase titania/poly(lactide-co-glycolide) composites of the highest nanometer surface roughness. Journal of Biomedical Materials Research 2006; 78A(4):798-807.

2 Traykova T, Aparicio C, Ginebra MP, Planell JA. Bioceramics as nanomaterials. Nanomed. 2006; 1(1):91-106.

3 Webster TJ, Siegel RW, Bizios R. Osteoblast adhesion on nanophase ceramics. Biomaterials 1999; 20(13):1221-1227.

4 Webster TJ, Ergun C, Doremus RH, Siegel RW, Bizios R. Enhanced functions of osteoblasts on nanophase ceramics. Biomaterials 2000; 21(17):1803-1810.

5 Fung YC. Biomechanics: Mechanical Properties of Living Tissues. New York: Springer-Verlag, pp. 500-519, 1993.

Topographcial Effects of Micro/Nano Crystallized Diamond for Controlling Osteoblast Adhesion

Lei Yang [1,a], Thomas J. Webster [1,2,b] and Brian W. Sheldon [1,c]
[1] Division of Engineering, Brown University, Providence, RI 02912, USA
[2] Department of Orthopedics, Brown University, Providence, RI 02912, USA
Email: [a] Lei_Yang@brown.edu; [b] Thomas_Webster@brown.edu; [c] Brian_Sheldon@brown.edu

Abstract:
Recent research has shown that diamond is a promising orthopedic biomaterial due to its excellent tribological and biocompatibility properties. Our previous research revealed that the adhesion of osteoblasts (bone forming cells) varies on different types of diamond surfaces. An approach to promote osteoblast functions by manipulating diamond topography is introduced in the present work. Three kinds of diamond with different crystallinity and microstructure (i.e., micron-crystallized diamond (MD), nano-crystallized diamond with spherical grains (NDs) and a mixture of diamond and amorphous carbon with nano platelet grains (NDp)) were fabricated and studied here. Topographical features of diamond were determined by Atomic Force Microscopy (AFM) and Scanning Electron Microscopy (SEM) and indicated that factors such as roughness, geometric and projected area ratios, and local height gradients contributed to altered contact angles and osteoblast adhesion. In summary, the results of this *in vitro* study suggested that nano diamond (if created appropriately) could promote (such as at the hip implant stem) or inhibit (such as at articulating surface) osteoblast adhesion necessary for improving orthopedic implant efficacy.

Keywords: Osteoblast adhesion; topography; nanocrystalline diamond

1. Introduction

Synthetic materials are now used, with differing degree of success, to replace failed organs in many biological systems (such as joints) [1]. In order to extend service life and relieve pain suffered by patients, artificial hip joints based mainly on alloys and ultrahigh molecular weight polymers are used after severe hip bone fracture. However, current hip implants have functional lifetime less than 10-15 years. Hip joint replacement requires extraordinary anti-wear and anti-corrosion properties as well as high strength and biocompatibility. New diamond coatings on metallic materials offers a practical solution to this problem because of their superior properties including extreme hardness [2], low friction coefficient [3], ultra-low wear rate [4], high chemical resistance [5], high fracture toughness [6] and biocompatibility [7]. Contrary to its well-known mechanical and tribological behaviors, the biological response (or more accurately, osteoblast cellular response) to diamond has not been extensively investigated to date. Recent work focusing on cellular adhesion (or proliferation) on diamond using different cell lines [7-10] provided preliminary quantity or morphological evidence of cells. Other work connected the hydrophobicity of diamond films to the adsorption of proteins involved in cellular adhesion but did not report cell behavior [11].

In this work, an attempt to correlate a relationship between the features of diamond that cells interact directly with (such as topography of diamond films) and corresponding osteoblast behavior is introduced. The investigation starts with the topographical control and characterization of micron/nano-crystalline diamonds produced by microwave plasma enhanced chemical-vapor-deposition (MPCVD) and ends in the resulting variation in osteoblast adhesion on different types of diamond. The connections between these two parts are established by contact angle measurements and the observation of the control of osteoblast adhesion (either promotion or inhibition) based on properties of the diamond coating.

2. Experimental

2.1. Materials

Diamond used in this work was produced by MPCVD. Different diamond films were deposited on a commercial silicon wafer (phosphorus doped, (100) face polished, Virginia Semiconductor, PA) with 0.6% methane, 5%~25% hydrogen and argon at 800°C and a chamber pressure of 111 torr. The silicon substrates were seeded with nano diamond powder of an average size of 50 nm before film growth. Growth time of all films was 2 hrs. A kind of special diamond film was grown without direct exposure to plasma was also investigated here.

2.2. Characterization of Diamond Topography and Composition

Surface morphology of diamond was observed by scanning electron microscopy (SEM, LEO 1530VP, Zeiss) at an accelerating voltage of 4 kV. Raman spectroscopy (Renishaw Ramascope) was applied to identify the chemical composition of the diamond films. Topography of the diamond films was further investigated by atomic force microscopy (AFM, AutoProbe, Park Scientific Instrument). Silicon nitride tips and contact mode were used at a scanning speed of 0.5 Hz. At least three regions on the same sample were selected for analysis of topographical parameters describing topography of diamond films.

2.3. Measurement of Contact Angles

Contact angles on diamond surfaces were measured through the sessile drop shape method. 10 μL of deionized water was used to form a drop and its shape was recorded by the drop shape analyzer (Easy Drop FM40, Kruss, Germany). Contact angles were then calculated by computer software. At least five spots on each sample were selected for measurements. Contact angles of silicon wafer were also measured for comparison purpose.

2.4. Osteoblast Adhesion Test

Human osteoblast (CRL-11372, ATCC) of population number between 10 and 15 was cultured in Dulbecco's Modified Eagle's Medium (DMEM) until they reach 80~100% confluence, and then were used for adhesion tests. The diamond films were cleaned by soaking in a series of solutions including acetone, 100% ethanol, 70% ethanol and dionized water. Diamond was first soaked in such solutions for 10 min. and then in ultrasonic cleaner for another 10 min. Then, the samples were dried at 80 °C and sterilized by UV light exposure overnight before cell culture. Silicon wafer for diamond film growth were used as a control.

The adhesion seeding density of osteoblast was 3500 cells/cm^2. After seeding on diamond, the cell culture wells were moved to CO_2 humidified incubator (5% CO_2, 37°C) for 4 hrs. Cells were then fixed with 10% Formalin (Fisher Scientific, Co.) and stained by DAPI (Sigma Aldrich Co.). The adherent cells were counted under a fluorescence microscope (FM, Axiovert 200M, Zeiss) and then osteoblast density was calculated.

Osteoblast distribution on diamond after their adhesion was observed under SEM (Leo 1530VP, Zeiss). Cells were fixed with formalin and then dried in the oven in order to expose the diamond substrate underneath the cell membrane.

3. Results and Discussion

3.1. Topography of Diamond Films of Varied Crystallite Size

The Topography of the diamond films varied dramatically with H_2 concentration in plasma. Table 1 summarizes the H_2 concentration used for film growth and resulting features of the diamond films. SEM and AFM images in Fig.1 showed the large variation of diamond crystallite and topography as the H_2 concentration increased. Nearly equal-axial, round nano crystallites were formed with H_2 concentrations blow 10%, and the crystallite size was a few tens of nanometers. As the H_2 concentration reached 15% and above, growth of nanocrystalline diamond resulted in the formation of large crystallographics facets (e.g. (111) and (100) plane) and consequently micron size tetrahedral or square shapes of diamond grains. The mechanism related to the evolution of grain size and shape in diamond films has been discussed somewhere [12].

Table 1 H_2 concentration for diamond film growth and roughness

Name*	H_2 concentration (% flow rate)	R_{pv} (nm)**	R_{rms} (nm)***
NDp	20	147.9	12.9±3.7
ND5	5	159.4	19.8±3.1
ND10	10	434.8	24.3±5.6
MD15	15	1535.0	76.2±2.4
MD20	20	887.9	63.9±4.1
MD25	25	965.4	70.8±3.5

* ND- nanocrystalline diamond, MD-microncrystalline diamond, NDp-grown without direct exposure to plasma; numbers correspond to H_2 concentration.

peak-valley roughness. * root mean square roughness.

Roughness results of diamond films obtained from AFM (Table 1) agree with the topographical features shown in Fig.1. Root mean square roughness (R_{rms}) of diamond films increased gradually as more H_2 was added, which was explicit because the grains on the surface grew bigger and made the surface rougher. However, the variation of peak-valley roughness (R_{pv}) revealed an exception on MD15, which has a comparatively smaller R_{rms} but the highest R_{pv}. This attributes to the definition of R_{rms} in that it is a statistical calculation of surface height. Thus, few but large 'spikes' on the surface (implies a large R_{pv}) can also result in a small R_{rms} value. This suggests that R_{rms} may not be a perfect parameter for the determination of roughness, or at least it should not be the sole parameter to describe surface topography.

Topographical features of a special diamond film (the NDp) are also shown in Fig.1f and Table 1. NDp reveals the topography composed of nano size platelets. The platelets are in irregular shape on the edges and some sharp, thorn like edges were observed. NDp was fabricated in high H_2 concentration (20%) but not exposed to the plasma directly. The Raman spectra show that the NDp film consisted of diamond and amorphous carbon (spectrum is not shown).

The results above imply the possibility that topography features of diamond films, including but no limited to grain size and morphology as well as surface roughness, can be simply controlled by H_2 concentration during MPCVD.

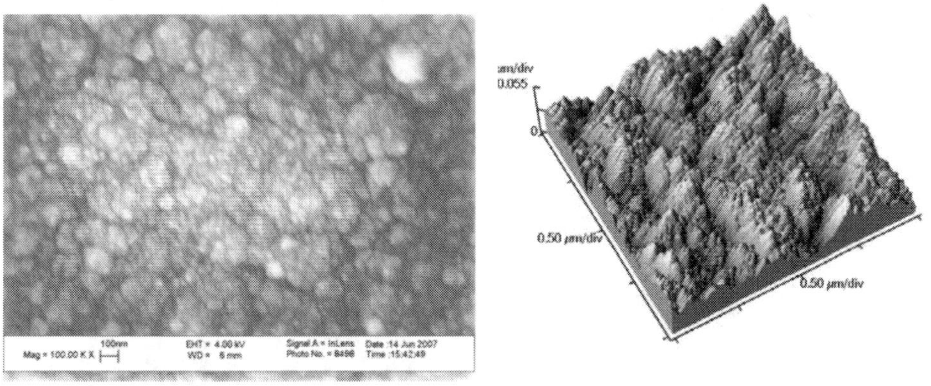

(a)

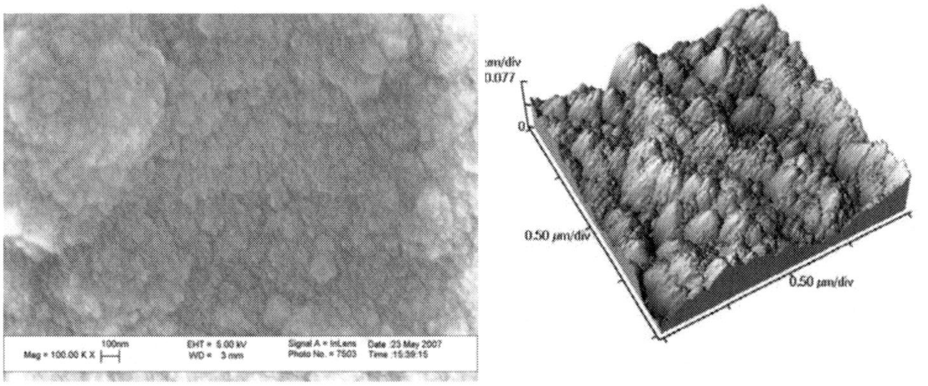

(b)

Fig. 1 SEM (left) and AFM (right) images of (a) ND5 and (b) ND10 (continued on next page).

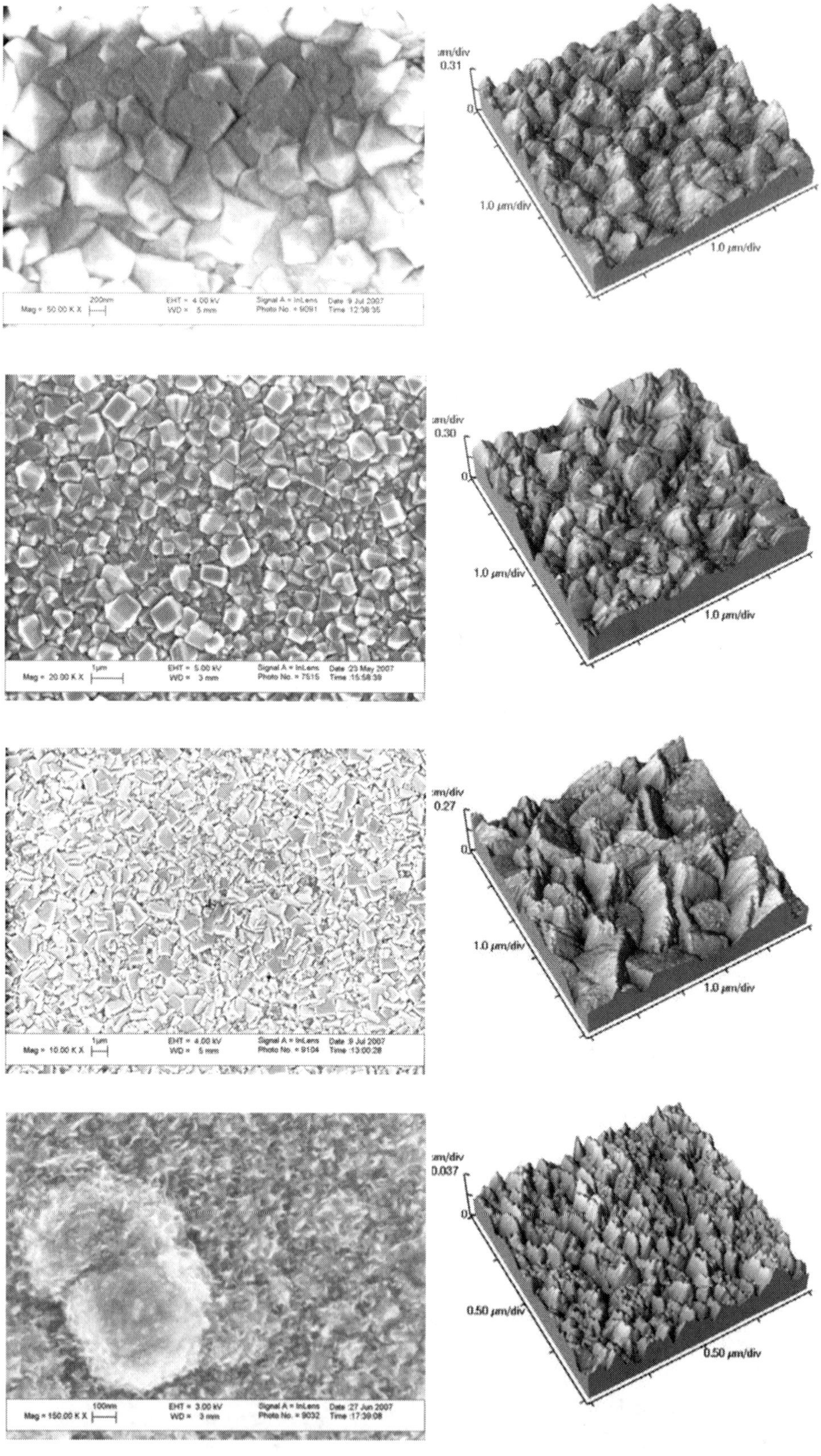

Fig. 1 (continued) SEM (left) and AFM (right) images of (c) MD15; (d) MD20; (e) MD25 and (f) NDp.

3.2. Topography Effects on Contact Angles of Diamond Films

Contact angles can provide information of surface wettability (hydrophobicity or hydrophilicity) and surface energy (although not directly). Fig. 2 shows the contact angles measured on different diamond films. Results showed that micron-crystalline diamond films have greater contact angles than nano-crystalline ones. NDp had an even smaller contact angle than these two groups of diamond films.

The definition of contact angle gives:

$$\cos\theta = \frac{\gamma_{SV} - \gamma_{SL}}{\gamma_{LV}} \qquad (1)$$

where θ is contact angle, γ_{SV} is interfacial tension between the solid and vapor, γ_{SL} is the interfacial tension between the solid and liquid and γ_{LV} is the tension between liquid and vapor. Usually γ_{LV} is not related to materials and γ_{SV} doesn't vary too much in the present case because the bonding between water and diamond doesn't change if one assumes nano and micron diamond are chemically identical. Then, γ_{SV} becomes critical to determine the value of θ.

However, conventionally defined γ_{SV} assumes the solid surface is smooth, which is not true for real materials. Topographical features of rough surfaces can make the geometric surface area greater than the projected surface area [13]. If the geometric surface area is greater, then nominal surface energy that is averaged by the projected area, i.e.,γ_{SV}, will increase and as a result, θ increases.

Therefore, the contribution of topography effects to γ_{SV} is closely related to the ratio of the geometric surface area to the projected area. So, one can introduce a parameter, τ_{area}, to represent this ratio and to connect the topographical feature of diamond films to the contact angles:

$$\tau_{area} = \frac{A_{geo}}{A_{proj}} \qquad (2)$$

Where A_{geo} and A_{proj} are geometric areas and projected areas obtained from AFM data, respectively. Each τ_{area} of the diamond film is shown in Fig.2. τ_{area} of the silicon wafer was set as unity since it is very smooth. There is a very good correlation between τ_{area} and contact angles of each sample, i.e., diamond with high τ_{area} is more hydrophobic than the one with low τ_{area}. A similar result can be found elsewhere [13].

It is worth to point out that R_{rms} is usually positively correlated to τ_{area} too, but the discrepancy may occur just like the situation of R_{pv} and R_{rms} described above.

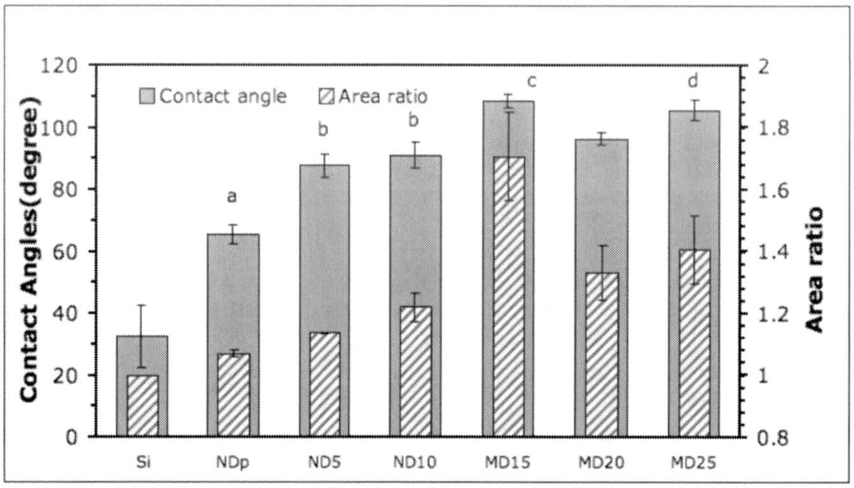

Fig. 2 Contact angles of the diamond films. Data=mean±SEM. Different alphabetical letters indicate statistically significant differences, $p<0.01$ between a and b, b and c; $p<0.05$ between b and d.

3.3. Topography Effects on Osteoblast Adhesion on Diamond Films

Osteoblast adhesion results are shown in Fig. 3. Nano-crystalline diamond shows enhanced cell adhesion, while micron-crytalline diamond and NDp revealed worse cell adhesion. For diamond films, it is not surprising that more hydrophilic films have better cell adhesion. One of the possible reasons for this is that proteins or molecules necessary for cell adhesion can easily move and absorb to a hydrophilic surface. For NDp that is more hydrophilic than NDs and MDs, the decrease in cell adhesion can be due to the different surface chemistries and platelet-like topography of the films. The cell adhesion difference on NDs and NDp can be applied to different parts of artificial hip joint. For instance, NDs can be coated on the stem of a hip implant to promote cell adhesion. In contrast, NDp may be optimal for the inside of acetabular articulation where cell adhesion is not preferred but wear-resistance is important.

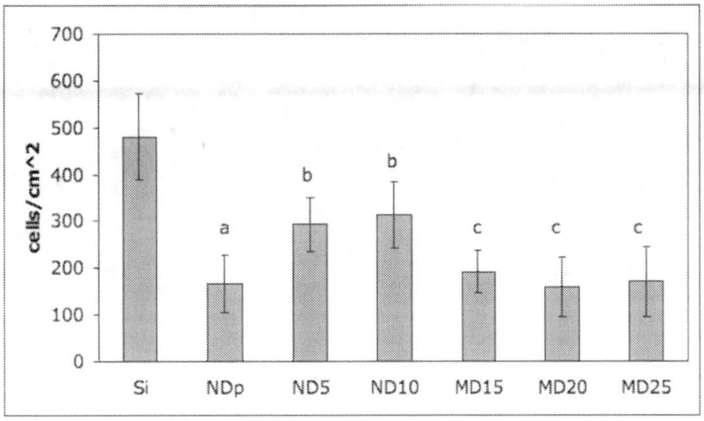

Fig. 3 Osteoblast adhesion on the diamond films. Data=mean mean±SEM. Different alphabetical letters indicate statistically significant differences, $p<0.01$ between a and b, $p<0.05$ between b and c.

A further observation on osteoblast distribution on the diamond films created in the study is shown in Fig. 4. The diamond sample was MD25. Osteoblast extended its membrane radially on the diamond surface and its filopodia extended as far as tens of microns away from its nuclei (Fig. 4a). A close observation of one of the filopodia (Fig. 4b) reveals that it extended along paths of small and continuous profile height gradients (indicated by red arrows). The variation of local topography changed as "slow" as possible in this region. Interestingly, osteoblast filopodia at the positions indicated by blue arrows chose a much rougher path rather than the smooth crystal facets to extend, indicating a path with large height drops (reasonable for micron size crystal facet) is not favorable by filopodia. The relationship of τ_{area} and hydrophobicity described earlier can explain this phenomenon. A large height gradient on diamond surfaces indicates a high local R_{pv}, or high τ_{area} and possibly high R_{rms}. Thus, this kind of localized position on the film can be more hydrophobic than others, which is not favorable by cells. These results provide evidence that topography features of diamond films, including surface roughness, local height gradient or τ_{area}, affect osteoblast adhesion.

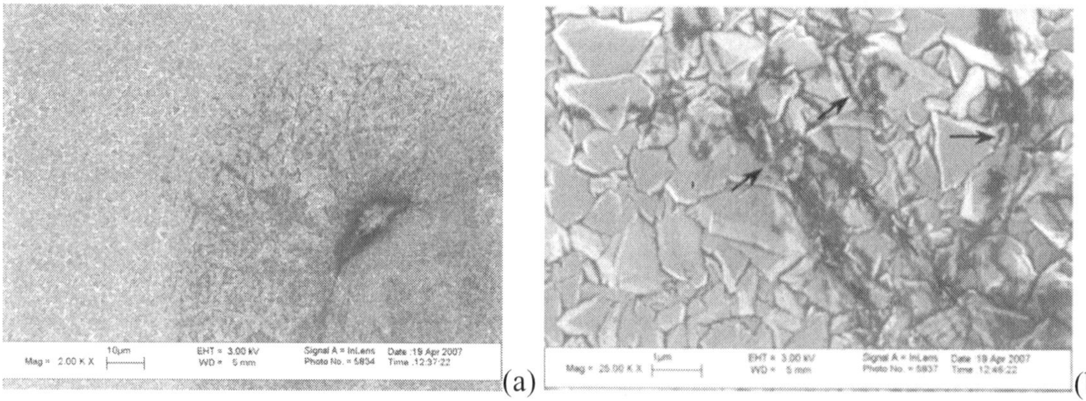

(a) (b)

Fig. 4 (a) Osteoblast distribution on the diamond film MD25; (b) distribution of osteoblast filopodia on the diamond film MD25.

4. Conclusions

Topographical features of diamond film (such as roughness R_{pv}, R_{rms}, geometric and projected area ratio τ_{area}, and local height gradients) contributed to altered contact angles and osteoblast adhesion. The results of this study suggested that diamond films (if created appropriately) can promote or inhibit osteoblast adhesion necessary for improving orthopedic implant efficacy.

6. References

[1] George Whitesides, Amy Wong. The intersection of biology and materials science. MRS Bulletin, 31(2006)1: 19-27.

[2] ME O'Hera, CJ Clausing, WC Oliver, RH Parrish. Determination of mechanical properties of diamond and diamond-like films by the ultralow load indentation techniques. Mater Res Soc Extended Abstr 1989; EA-19:131-137.

[3] IP Hayward, IL Singer. The tribological behavior of diamond coatings. In: Messier R, Glass JT, Bulter JE, Roy R, eds. Proc 2nd Int Conf New Diamond Sci Technol. Pittsburgh, PA: Materials Research Society, 1986:785-789.

[4] Lappalainen R et al, J.Biomed.Mater.Res, 66B (2003): 410-413.

[5] KE Spear. Diamond-ceramic coating of the future. J. Am Ceram Soc 72(1989): 171-191.

[6] MD Drory, CF Gardinier, JS Speck. Fracture toughness of chemically vapor-deposited diamond. J Am Ceram Soc. 74(1991): 3148-3150

[7] L Tang, C Tsai, WW Gerberich, et al. Biocompatibility of chemical-vapor-deposited diamond. Biomaterials 16 (1995): 483-488.

[8] SE Rodil, R Olivares, H Arzate, S Muhl. Properties of carbon film and their biocompatibility using in-vitro tests. Diamond and Related Materials, 12 (2003): 931-937.

[9] TV Kumari, PR Anil Kumar,, et al. In vitro cytocompatibility studies of diamond like carbon coatings on titanium. Bio-medical Materials and Engineering, 12(2002):329-338.

[10] C Du, XW Su, FZ Cui, XD Zhu. Morphological behavior of osteoblasts on diamond-like carbon coating and amorphous C-N film in organ culture. Biomaterials, 19(1998)651-658.

[11] JM Garguilo, BA Davis, M Buddie, FAM Kock, RJ Nemanich. Fibrinogen adsorption onto microwave plasma chemical vapor deposited diamond films. Diamond and Related Materials, 13 (2004): 595-599.

[12] D Zhou, DM Gruen, LC Qin, et al. Control of diamond film microstructure by Ar additions to CH_4/H_2 microwave plasma. J Appl. Phys. 84 (1998) 4: 1981-1989.

[13] HY Erbil, AL Demirel, Y Avcı, O Mert. Transformation of a simple plastic into a superhydrophobic surface. Science, 299 (2003): 1377-1379.

134e Biofunctionalized Albumin Nanoparticle-Polymer Composites For Skin Tissue Engineering And Wound Remodeling

Maria Pia Rossi, Department of Chemical Engineering, NJ Center for Biomaterials and Rutgers University, 599 Taylor Road, Piscataway, NJ 08854, Ram I. Sharma, Chemical and Biochemical Engineering, Rutgers University, 98 Brett Road Room C-230, Room C-001, Piscataway, NJ 08854, Vanesa Figueroa-Tanon, Cellular and Molecular Biology, Rutgers University, 190 Frelinghuysen Road, Piscataway, NJ 08854, Joachim Kohn, Chemistry and Chemical Biology, Rutgers University and The New Jersey Center for Biomaterials, 145 Bevier Road, Piscataway, NJ 08854, Jean E. Schwarzbauer, Molecular Biology, Princeton University, Schultz Lab-315, Princeton, NJ 08544, and Prabhas V. Moghe, Rutgers University, 98 Brett Road, Piscataway, NJ 08854.

Nanoparticles synthesized from albumin, an innate, ubiquitous protein found in the human body, are currently being explored for applications such as drug or gene delivery due to their biocompatibility and biodegradability. We have functionalized albumin nanoparticles (ANPs) ranging from 40 to 100 nm with the recombinant fragment from fibronectin (FNf) comprised of the RGD-containing module 10 and the synergy-region-containing module 9 by conventional protein conjugation chemistry. When presented on two-dimensional substrates, the functionalized ANPs aided in the accelerated secretion and assembly of extracellular matrix proteins by fibroblasts; additionally, they promoted activated migration by keratinocytes, two processes that are crucial during wound healing. The functionalized ANPs also have been presented on electrospun, three-dimensional poly(DTE carbonate) scaffolds that are bioresorbable and biodegradable. The scaffold-ANP composites were then seeded with fibroblasts and keratinocytes to study the possible establishment of a three-dimensional (3-D) skin analog. The nanoparticle-decorated, 3-D scaffolds provide an effective platform for the study of tree-dimensional migration of fibroblasts into the scaffolds and extracellular matrix protein secretion, including collagen and fibronectin production. Following prolonged fibroblast culture, seeding of keratinocytes onto the scaffold was performed to investigate keratinocyte "epithelialization" and stable adhesion onto the newly secreted matrix.

134f Characterization Of Whey Protein Isolate Sol-Gels As Scaffolds For Bone Regeneration

Mia Dvora and James E. Henry. Cain Department of Chemical Engineering, Louisiana State University, 110 Chemical Engineering, South Stadium Rd, Baton Rouge, LA 70803

Conventional alternatives used for bone replacement, such as metals, ceramics, and various polymers have all shown considerable shortcomings. The deficiencies are due, in part, to the inability of these materials to fulfill the nonstructural roles of bone, such as ion reserves and red and white blood cell production. To further complicate matters, metal implantation generally results in necrosis of the adjacent bones and ligaments. Furthermore, the metal does not allow for the regeneration of bone in the site of the implant. While ceramics overcome the biocompatibility issues of metals, ceramics do not degrade and show high fragility and, therefore, cannot be used for large implants. Polymers are being investigated to address many of these issues. The chemistry in their construction can be used to control their rate of degradation and cross-linking can be used to increase the compressive properties of the material. However, developing polymers that meet all the requirements for the scaffold material has proven difficult. With hopes to correct many of these deficiencies, we are investigating using whey protein as a scaffold material.

The proposed scaffold material is a sol-gel made by thermally treating a concentrated suspension of whey protein isolate in an aqueous calcium chloride solution. Two aspects of the potential scaffolds are characterized in this study. The first is the material strength and its dependence on the protein and calcium chloride content of the gel-forming suspension. The second is the affinity of preosteoblastic cells towards the gels, and its dependence on the same two variables. As osteointegration is a key factor for the success of a scaffold material, the compositional range of interest for future work is to be gathered from this study. The significance of the work presented is that the strength of the naïve material is comparable to that of bone, and that the affinity of cells to the scaffold is such that its presence encourages osteoconduction. Both characteristics make this a promising future implant material in bone-deficient patients.

First, compressive material testing was performed on gel samples comprising a range of protein concentrations and calcium chloride concentrations to determine the ultimate compressive strength and modulus of the material. In this way the composition of highest strength is found and the range of interest for further study is determined. On samples of this range, preosteoblastic mouse cell-line (MC3T3-E1 subclone 4), is cultured in α-MEM for various incubation periods. SEM micrographs of the fixed cells will be presented with an analysis of "flatness," or, degree to which the cells spread on the different samples. Along with these, viability and cell density results from flow cytometry will be presented, quantifying the growth kinetics of the cells on the proposed scaffold material.

153b Multifunctional Micro- and Nanoparticles Made by Electrified Co-Jetting

Joerg Lahann, University of Michigan, Dept. of Chemical Engineering, 3074 H.H. Dow Bldg., Ann Arbor, MI 48109-2136

Recent advances with particle designs have generated increasing enthusiasm for use of polymer particles in drug delivery or as imaging probes. Because of their controllable sizes, surface characteristics, and degradation rates, polymer particles may carry potential for mimicking biological systems, such as viruses or circulating blood cells. While a series of so-called "multifunctional particles" has been prepared already, individual particle properties are typically convoluted and often cannot be controlled independently.

We have recently developed a disruptive technological approach to fabricate anisotropic particles with independently controllable compartments [1-4]. The fabrication process relies on electrified co-jetting, which is an off-spring of the electrospinning technology. Using co-extrusion of multiple fluid jets, manipulation in an electrical field is used to fabricate multiphasic particles. The geometry of the multi-compartment particles is induced by sustaining the interfaces between the two or more jetting solutions during the jet fragmentation and size reduction accompanying solvent evaporation. Therefore, the individual compartments can be separately loaded with biomolecules or surface-modified with reactive bioligands. Moreover, tight control of process parameters allows for controlling the size and shape of the multiphasic micro- and nanoparticles. In such nanomaterials, the anisotropic distribution of the two or more different chemical species becomes another governing rule – in addition to size and shape. In principle, such novel particle geometries enable independent control of key parameters, such as chemical composition, surface functionalization, biological loading, shape, and size to be controlled independently for each compartment – thereby effectively mimicking the architecture of their biological counterparts, i.e., cells or viruses.

The co-jetting technology lends itself to fabrication of diverse compartment geometries. For instance, biphasic particles with equally sized compartments can be prepared. Similarly, particles with more than two compartments can be fabricated. Modification of process parameters can yield in biphasic particles, which have core/shell geometries. At the same time, compartments can be made of different matrix materials, which enable independent surface modification, as well as independent control of extremely diverse functionalities, such as release rate, environmental responsiveness, or color, to name just a few.

[1] K. H. Roh, D. C. Martin, J. Lahann, Nat. Mater. 2005, 4, 759. [2] K. H. Roh, M. Yoshida, J. Lahann, Langmuir 2007, in press. [3] M. Yoshida, K. H. Roh, J. Lahann, Biomaterials 2007, in press. [4] K. H. Roh, D. C. Martin, J. Lahann, J. Am. Chem. Soc. 2006, 128, 6796.

153c Engineering Dynamics of Nanoscale Biointerfaces for Enhanced Cell Motility and Matrix Assembly

Prabhas V. Moghe, Chemical and Biochemical Engineering; Biomedical Engineering, Rutgers University, 599 Taylor Road, Piscataway, NJ 08854

The ability of engineered bioactive materials to dynamically guide tissue repair and regeneration remains a challenge. Cells of the skin in particular engage in highly dynamic interactions with their matrix, which offers a platform to engineer nanoscale matrix display and modulate cell behaviors important for re-epithelialization and matrix remodeling during wound repair.

We examined the possibility of engineering dynamic biointerfaces, through the display of cell-interactive biological ligands from nanoscale albumin-based particle-substrates that would permit dynamic cell-material interactions following cell adhesion, including nanoparticle endocytosis. The underlying hypothesis is that matrix adhesion ligands can concertedly "superactivate" cells through dynamic ligand fates (mobility, cytointernalization) beyond ligand clustering and ligand-receptor binding. Features of the underlying ligand carriers can be used to systematically engineer these dynamics.

We report that over a certain regimen, the size scale of such ligand carriers can indeed accelerate ligand clearance dynamics, leading to efficient cellular processes such as motility. In contrast, different scale regimen control other cell behaviors, such as cell contractility and matrix assembly. Ongoing challenges include elucidation of the mechanisms of biological signaling and tracking intracellular single nanoparticle dynamics. Opportunities for tissue engineering and wound repair therapies include developing dynamic biointerfaces for navigating cells in 2-D as well as in 3-D.

187a Advances in Biosensing Methods

A. Sadana[1], Reema Taneja[2], and Kennon C. Shelton[2]. (1) University of Mississippi, Rebel Drive, University, MS 38655-1848, (2) Chemical Engineering Department, University of Mississippi, Rebel Drive, University, MS 38677-1848

A fractal analysis is presented for the binding and dissociation (if applicable) kinetics of analyte-receptor reactions occurring on biosensor surfaces. The applications of the biosensors have appeared in the recent literature. The examples presented together provide the reader with a perspective of the advances in biosensors that are being used to detect analytes of interest. This should stimulate interest in applying biosensors to other areas of application. The fractal analysis limits the evaluation of the rate constants for binding and dissociation (if applicable) for the analyte-receptor reactions occurring on biosensor surfaces. The fractal dimension provides a quantitative measure of the degree of heterogeneity on the biosensor surface. Predictive relations are presented that relate the binding rate coefficient with the degree of heterogeneity or the fractal dimension on the biosensor surface. The biosensor systems analyzed include (a) the binding of different concentrations of mouse monoclonal anti-rabbit IgG in solution to rabbit patterned cells using a diffraction-based sandwich immunoassay (Goh et al., 2003), (b) the binding and hybridization of DNAs using a double-wavelength technique for surface plasmon resonance measurements (Zybin et al., 2005), binding and dissociation of GSH (Glutathione-MPC (monlayer protected clusters) in solution to polyclonal antibody plus protein A immobilized on a quartz crystal microbalance (QCM) (Gerdon et al., 2005), and the (d) binding and dissociation of substance P (neuropeptide) (SP) in solution to quencher-labeled MAb (monoclonal antibody) SP31 using SPIT-FRI (solid-phase immobilized tripod for fluorescent renewable immunoassay) procedure (Volland et al., 2005).

Glucose Biosensor Using Non-Woven Single-Walled Carbon Nanotube Films

*Wenzhao Jia[1],Jiangxiong Wang[2], Xiaowei Sun[2], Li Song[3], Sishen Xie[3] and Yu Lei[1] **

[1] Department of Chemical, Materials and Biomolecular Engineering, University of Connecticut, Storrs, CT 06269-3222

[2] School of Electrical and Electronic Engineering, Nanyang Technological University, Nanyang Avenue, Singapore 639798

[3] Institute of Physics, Center for Condensed Matter Physics, Chinese Academy of Sciences, Beijing, P. R. China, 410082

*Corresponding author email: ylei@engr.uconn.edu

ABSTRACT

A glucose biosensor, based on detecting the reduction of gluconolactone and hydrogen peroxide, has been developed utilizing glucose oxidase and non-woven single-walled carbon nanotube (SWNT) films synthesized by floating chemical vapor deposition technique. The strongly catalytic reduction of gluconolactone and hydrogen peroxide generated from the enzymatic reaction of glucose was observed at applied potential of -0.1 V (vs. Ag/AgCl) and led to the selective detection of glucose with a high sensitivity of 34.375 $\mu A/cm^2 \cdot mM$, a fast response time of less than 15 s and a limit of detection (LOD) of 0.05 mM. At such low applied potential, the interference from common interfering agents such as uric acid and ascorbic acid can be minimized. These results demonstrate that the non-woven SWNT films have potential applications in constructing highly sensitive, stable and fast response biosensors.

KEYWORDS: SWNT; glucose; biosensor; glucose oxidase; amperometric.

1. Introduction

Diabetes is a metabolic disorder and a major world health problem. There are over 170 million diabetics worldwide (WHO 2004) and the number is projected to 300 million in 2025, so glucose biosensors are becoming incredibly important.[1,2] Due to its high selectivity to glucose and high activity over a broad range of pH,[3] glucose oxidase (GOx), which can catalyze glucose to generate hydrogen peroxide and gluconolactone, has been widely used to construct various amperometric glucose biosensors based on the detection of hydrogen peroxide.[1,4-7]

Since carbon nanotube (CNT) has been discovered by Iijima in 1991,[8] multi-walled carbon nanotube (MWNT) and single-walled carbon nanotube (SWNT) have attracted much attention in the biosensor development due to high surface-to-volume ratio, high electrocatalytic capability and fast electron transfer rate.[9-15] Recently, there has been a growing interest in applying SWNTs in the development of biosensors.[1,9] Methods such as depositing SWNT on a conducting electrode or mixing SWNT in an inert matrix to form a paste electrode[11,16-19] have been widely used to prepare SWNT modified electrode, however, biosensor fabricated using these methods are mechanically unstable and their fabrication processes are usually tedious,[20,21] limiting their further application.

In this work, we report a glucose biosensor using non-woven SWNT film synthesized directly by chemical vapor deposition (CVD) and GOx as the sensing element. The SWNT film shows excellent mechanical properties and can be used as nanomaterials for GOx immobilization. The catalytic reduction of hydrogen peroxide and gluconolactone generated from the enzymatic reaction was observed at applied potential of -0.1 V (vs. Ag/AgCl) and led to the selective detection of glucose with a high sensitivity of 34.375 μA/cm^2•mM, a fast response time of less than 15 s and a limit of detection (LOD) of 0.05 mM. At such low applied potential, the interference from common interfering agents such as uric acid and ascorbic acid can be minimized. These results indicate that the non-woven SWNT film is a promising material for constructing biosensors.

2. Experimental Section

GOx (190 units/mg) and Nafion solution (5 wt%) were purchased from Sigma-Aldrich. D-(+)-glucose was obtained from Merck. Sodium di-hydrogen phosphate and di-sodium hydrogen phosphate were bought from Panreac and VWR respectively.

The non-woven SWNT films were synthesized by floating CVD method using ferrocene as catalyst in a quartz tube. The details of the synthesis method of SWNT films have been described elsewhere.[22] After the growth, the non-woven SWNT films with large area were peeled off from the wall of the quartz tube and ready for use. Field emission scanning electron microscopy (FESEM; JSM-6500F) and high-resolution transmission electron microscopy (HRTEM; JEM-2100F) were employed in characterizing the SWNT films.

The as-prepared SWNT film was cut to small piece (4mm × 2 mm) using scissor and carefully spread on a glassy carbon electrode (GC) by dropping 50 μl ethanol and dried at 60 °C in an oven. Subsequently, a 5 μl 0.5 wt% nafion solution containing 15 mg/ml GOx (190 units/mg) was dropped onto the surface of SWNT film modified GC electrode and dried at room temperature for 50 min. The Gox-SWNT film-modified electrode was washed with deionized water and stored at 4 °C until use. The electrochemical experiment was performed at room temperature utilizing an electrochemical workstation (Bio-logic SA VMP2) with a conventional three-electrode configuration consisting of GOx-SWNT film modified GC working electrode, a platinum auxiliary electrode and an Ag/AgCl reference electrode. All experiments were carried out in a glass beaker containing 30 ml of 0.01 M phosphate buffer (pH 7.4) at room temperature.

3. Results and Discussion

Figure 1A shows an optical image of the as-prepared SWNT film. The diameter of the SWNT film can reach several centimeters. The SWNT film shows excellent mechanical properties, and can be handled and manipulated easily using tweezers and scissors. A typical SEM image of the pristine SWNT film is shown in Figure 1B. It can be seen from Figure 1B that, the porous SWNT film is consisted of many long CNT bundles decorated with small diameter nanoparticles. Meanwhile, the CNT bundles entangle each other and form a dense network. Figure 1C is a HRTEM image of the SWNT bundles in the film. The HRTEM image reveals that the carbon nanotube diameter is of about 1~2 nm, which indicates that the film consists of single wall carbon nanotubes. The presence of catalyst residues was indicated with the shadow area in the HRTEM image. EDX element analysis result was shown in Figure 1D. The peak at 0.28 keV is related to C and the peaks at 0.71, 6.40 and 7.06 keV associated to Fe, which match our expectation because SWNT film is synthesized using ferrocene as catalyst.

The biosensor response to the addition of glucose at different applied potentials (vs. Ag/AgCl) is illustrated in Figure 2. At the applied potentials of 0 V and -0.1 V, SWNT film based glucose biosensor generated a significant reduction current in the step-down format whereas a step-up current was observed at the applied potentials of -0.2 V and -0.3 V respectively (Figure 2). This phenomenon can be interpreted as follows. The immobilized GOx catalyzes glucose to release hydrogen peroxide (H_2O_2) and gluconolactone. There are a few reports about the reduction of H_2O_2 recently, [1,9] however, according to the best of our knowledge, there is no report about the fate of gluconolactone, another product of enzymatic reaction, on the surface of GOx modified electrode. Gluconolactone can be hydrolyzed to gluconic acid spontaneously and gradually, but there is no obvious response observed by the addition of gluconic acid under the operating potential (Figure 3A). Our result (shown in Figure 3A) indicates that gluconolactone can be reduced on the surface of as-prepared SWNT film with a fast response of less than 15 s. As the reduction rate of gluconolactone on the SWNT surface is much faster than its spontaneous hydrolysis rate, it can be concluded that the observed reduction current upon the addition of glucose is contributed not only by H_2O_2 but also by gluconolactone, whereas the contribution of H_2O_2 is greater than that of gluconolactone. At the applied potentials of 0 and -0.1 V, the oxygen reduction is negligible and the observed reduction current is attributed to the reduction of H_2O_2 and gluconolactone liberated from the enzymatic reaction. However, at the applied potentials of -0.2 and -0.3 V, oxygen starts to be reduced at CNT surface.[23] As the oxygen is consumed by the enzymatic reaction and its concentration decreases in the solution, the oxygen reduction current has the step-up format in our detection system. The observed current at -0.2 and -0.3 V is the net effect of oxygen reduction and the reduction of H_2O_2 and gluconolactone. In order to maximize the biosensor response, limit the oxygen reduction and minimize the noise, -0.1 V (vs. Ag/AgCl) was selected for the detection of glucose subsequently.

Figure 3A displays a typical amperometric response of the SWNT film-based glucose biosensor after the successive addition of 0.5 mM and 1 mM glucose to the phosphate buffer (pH 7.4) under stirring. It can be seen from the plot that the biosensor showed a rapid and sensitive response to the change of glucose concentration with the response time of less than 15 s. The corresponding calibration curve of the SWNT film glucose biosensor is shown in Figure 3B. Following the increase of the glucose concentration, the reduction current increases and saturates at a high glucose concentration which might be attributed to saturation of enzymatic reaction. The calibration plot is linear up to 2 mM of glucose (R=0.9967) with a high sensitivity of 34.375 $\mu A/cm^2 \bullet mM$ and a limit of detection (LOD) of 50 μM (inset in Figure 3A). The sensitivity is significantly higher than those of glucose biosensors reported previously,[24, 25] which is attributed to the high surface-to-volume ratios, high electrocatalytic capability and fast electron transfer rate of SWNT film.

The selectivity of the biosensor was also investigated against normally co-existed interfering species (0.1 mM ascorbic acid and 0.1 mM uric acid) with glucose. As seen from Figure 3A, the interference from uric acid and ascorbic acid is negligible, which indicates good selectivity of the biosensor. The high selectivity can be attributed to the strong reduction of H_2O_2 and gluconolactone on SWNT film at low applied potential.

4. Conclusions

In conclusion, we have successfully demonstrated a SWNT film-based glucose biosensor for sensitive and selective determination of glucose. The strongly catalytic reduction upon the glucose addition was observed at an applied potential of -0.1 V (vs. Ag/AgCl) and led to the sensitive and selective detection of glucose. The observed current is attributed to the reduction of both gluconolactone and H_2O_2 on SWNTs. At such low applied potential, the interference from common interfering agents such as ascorbic acid and uric acid can be minimized. Additionally, the large area SWNT films used in the research have many unique advantages such as easy manipulation, excellent mechanical and conductive properties, biocompatibility and chemical inertness. All these features make non-woven SWNT film an attractive material in developing biosensors.

Acknowledgment

We greatly appreciate the funding from NSF (CMMI-0730826) and UConn large faculty research grant.

Figure captions

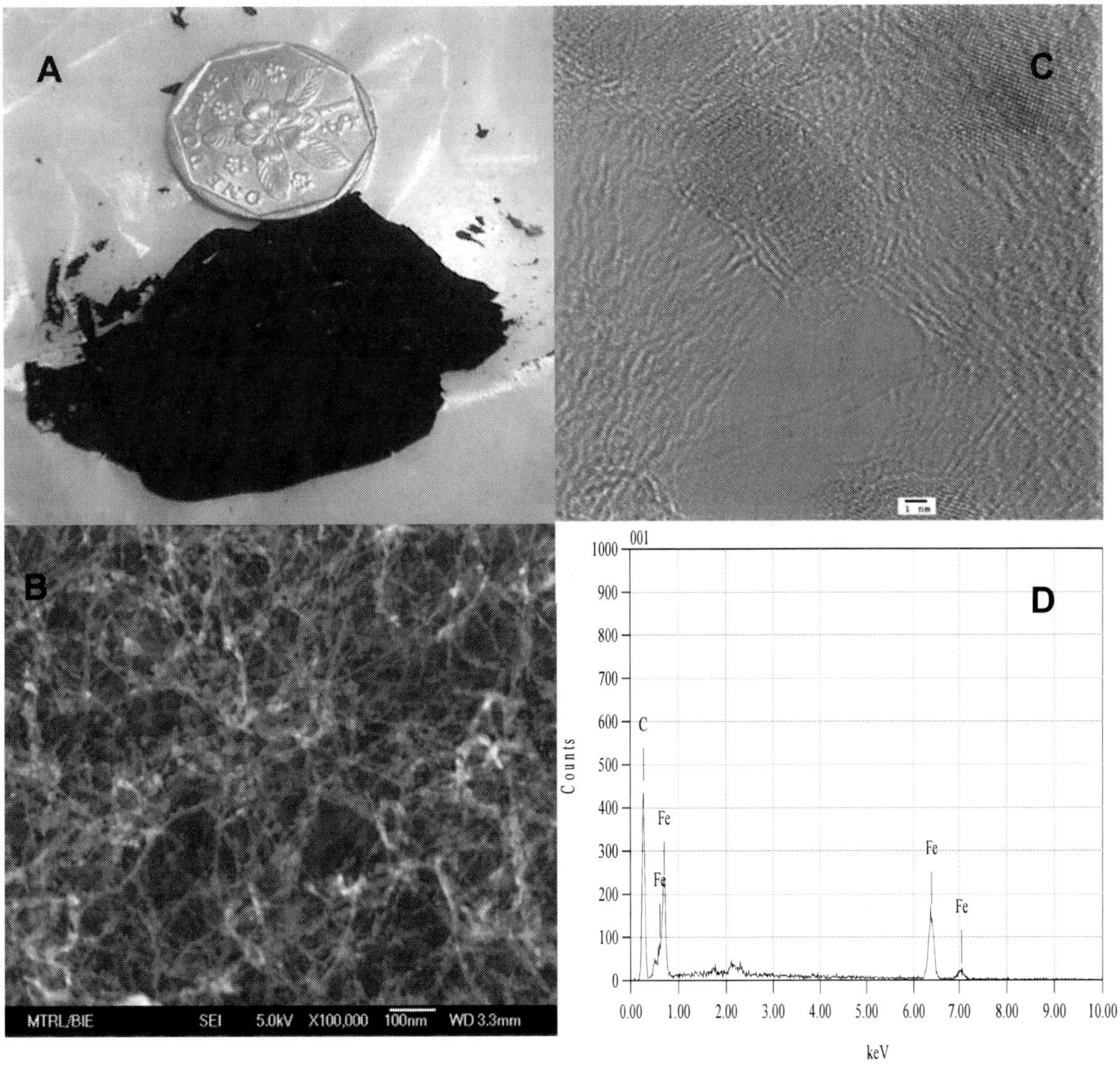

Figure 1. A) Optical image of an as-prepared non-woven SWNT film. B) SEM image of SWNT film. C) TEM image of some SWNT bundles in the SWNT film. D) EDX element analysis results.

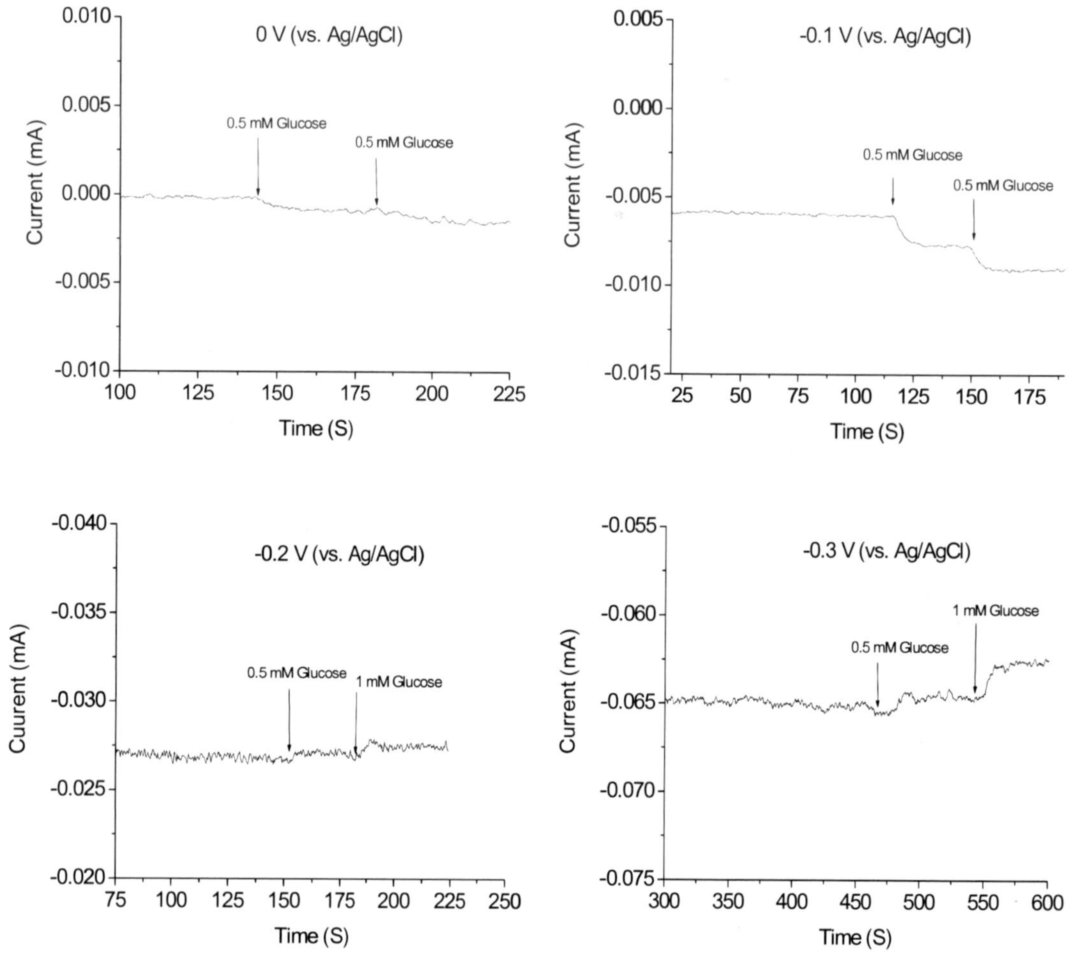

Figure 2. Biosensor response to the addition of glucose at different applied potential (vs. Ag/AgCl) in 0.01 M pH 7.4 phosphate buffer under stirring.

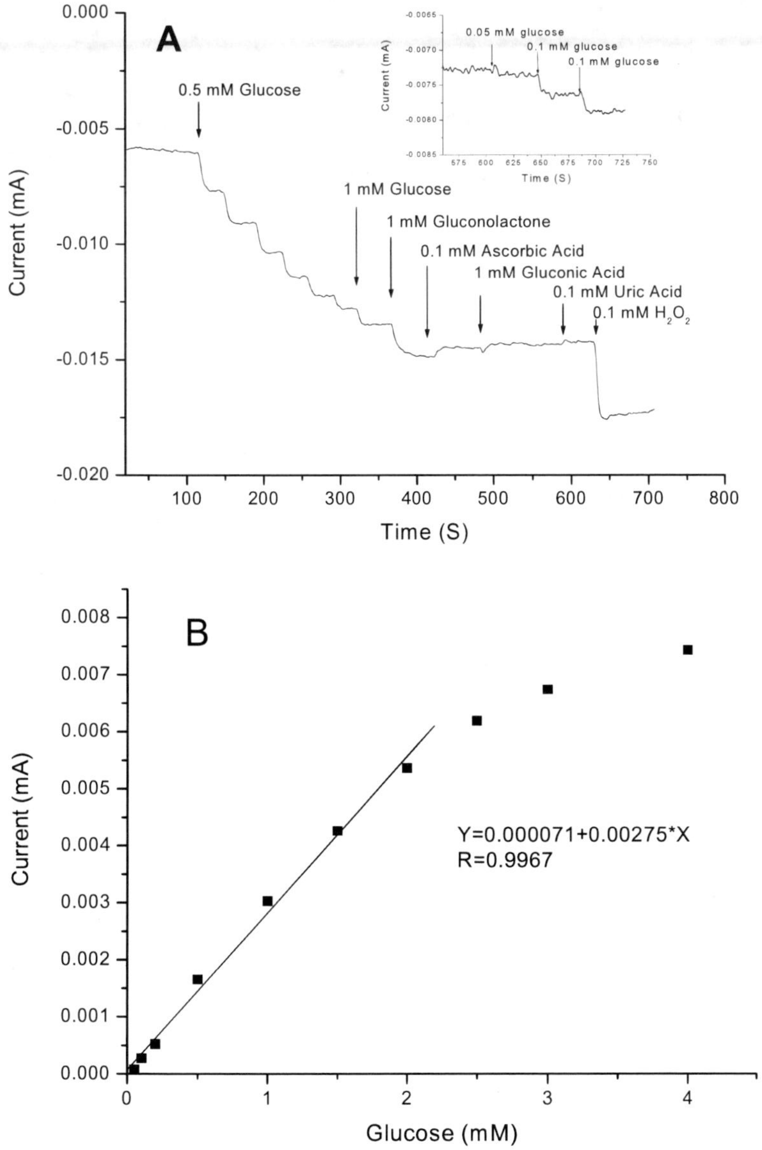

Figure 3. A) Amperometric responses of GOx-SWNT film modified GC electrodes with the addition of glucose, gluconolactone, ascorbic acid, gluconic acid, uric acid and H_2O_2 to the 0.01 M phosphate buffer (pH 7.4) under stirring. The applied potential was -0.1 V vs. Ag/AgCl reference electrode. The inset shows response for successive additions of 0.05 mM and 0.1 mM glucose, which represents experimental limit of detection. B) The corresponding calibration curve of SWNT film-based glucose biosensor at applied potential of -0.1 V (vs. Ag/AgCl) and operated in 0.01 M pH 7.4 phosphate buffer.

References

(1) Lin, Y.H.; Lu, F.; Tu, Y. ; Ren, Z.F. *Nano Lett.* **2004**, *4*, 191-195.

(2) Reach, G.; Wilson, G.S. *Anal. Chem.* **1992**, *64*, 381A-386A.

(3) White, B.J.; Harmon, H.J. *Biochem. Biophys. Res. Commun.* **2002**, *296*, 1069-1071.

(4) Frew, J.E.; Hill, H.A. *Anal. Chem.* **1987**, *59*, 933A-944A.

(5) Morris, N.; Cardosi, M.; Birch, B., Turner, A.P. *Electroanalysis* **1992**, *4*, 1-9.

(6) Wang, J.; Liu, J.; Chen, L.; Lu, F. *Anal. Chem.* **1994**, *66*, 3600-3603.

(7) Lee, D.; Lee, J.; Kim, J.; Kim, J.; Kim, H.B.; Kim, B.; Kim, C.H.; Shin, C.H.; Kwak, J.H.; Dohnalkova, A.; Grate, J.W.; Hyeon, T.; Kim, H.S. *Adv. Mater.* **2005**, *17*, 2828-2833.

(8) Iijima, S. *Nature* **1991**, *354*, 56-58.

(9) Wang, J.; Musameh, M.; Lin, Y.H. *J. Am. Chem. Soc.* **2003**, *125*, 2408-2409.

(10) Wang, J.; Li, M.; Shi, Z.; Li, N.; Gu, Z. *Anal. Chem.* **2002**, *74*, 1993-1997.

(11) Wang, J.; Musameh, M. *Anal. Chem.* **2003**, *75*, 2075-2079.

(12) Li, J.; Ng, H.T.; Cassell, A.; Fan, W.; Chen, H.; Ye, Q.; Koehne, J.; Han, J.; Meyyappan, M. *Nano Lett.* **2003**, *3*, 597-602.

(13) Baughman, R. H.; Zakhidov, A.; de Heer, W. A. *Science* **2002**, *297*, 787-792.

(14) Tsai, Y.C.; Li, S.C.; Chen, J.M. *Langmuir* **2005**, *21*, 3653-3658.

(15) Guan, W.J.; Li, Y.; Chen, Y.Q.; Zhang, X.B.; Hu, G.Q. *Biosens. Bioelectron.* **2005**, *21*, 508-512.

(16) Wang, J. *Electroanal.* **2005**, *17*, 7-14.

(17) Luong, J. H.T.; Hrapovic, S.; Wang, D. *Electroanal.* **2005**, *17*, 47-53.

(18) Merkoc, A.; Pumera, M.; Llopis, X.; Perez, B.; Valle, M. D.; Alegret, S. *Trend. Anal. Chem.* **2005**, *24*, 826-838.

(19) Day, T. M.; Wilson, N. R.; Macpherson, J. V. *J. Am. Chem. Soc.* **2004**, *126*, 16724-16725.

(20) Yu, X.; Chattopadhyay, D.; Galeska, I.; Papadimitrakopoulos, F.; Rusling, J. F. *Electrochem. Commun.* **2003**, *5*, 408-411.

(21) Gooding, J. J.; Wibowo, R.; Liu, J. Q.; Yang, W. R.; Losic, D.; Orbons, S.; Mearns, F. J.; Shapter, J. G.; Hibbert, D. B. *J. Am. Chem. Soc.* **2003**, *125*, 9006-9007.

(22) Song, L.; Ci, L. J.; Li, Lv; Zhou, Z. P.; Yan, X. Q.; Liu, D. F.; Yuan, H. J.; Gao, Y.; Wang, J. X.; Liu, L. F.; Zhao, X. W.; Zhang, Z. X.; Dou, X. Y.; Zhou, W. Y.; Wang, G.; Xie, S. S. *Adv. Mater.* **2004**, *16*, 1529-1534.

(23) Dai, Y.Q.; Shiu, K.K. *Electroanalysis* **2004**, *16*, 1697-1703.

(24) Vidal, J. C.; Garcia, E.; Castillo, J. R. *Biosens. Bioelectron.* **1998**, *13*, 371-382.

(25) Yu, J. H.; Liu, S. Q.; Ju, H. X. *Biosens. Bioelectron.* **2003**, *19*, 401-409.

187c Chemistry For A Single Walled Carbon Nanotube Fluorescence Based Glucose Sensor

Paul W. Barone and Michael S. Strano. Chemical Engineering, Massachusetts Institute of Technology, Building 66, Cambridge, MA 02139

Single walled carbon nanotubes (SWNT) fluoresce from 900 to 1600 nm affording them unique advantages over conventional fluorophores for use in in vivo applications. However, one of the main challenges in harnessing these advantages is maintaining SWNT stability in solution while imparting the necessary analyte selectivity. We demonstrate that coating the nanotubes in a partially hydrophobic dextran polymer allows detection of glucose based on changes in SWNT fluorescence. The prospects and feasibility of using such a system in vivo are discussed and elaborated on by imaging of SWNT fluorescence through tissue.

187d Photopolymerization for Signal Amplification in the Detection of Biomolecular Recognition

Ryan R. Hansen, Hadley D. Sikes, and Christopher N. Bowman. Chemical and Biological Engineering, University of Colorado, 1111 Engineering Dr. Mail Stop 0424, Boulder, CO 80309

With the recent development of diagnostic biochips comes the potential of high-throughput, rapid genetic screening of targeted pathogens and disease biomarkers. A primary limitation associated with this technology involves the sequence specific identification of low copy number polynucleotides using clinically suitable detection instrumentation, often requiring target amplification with PCR-based methods or through signal amplification methods. Current amplification methods suffer from limitations associated with individual primer selection for each targeted biomarker, reliance on enzymatic reactions that convert a substrate into a chemiluminescent or fluorescently detectable moiety, or dependence on sophisticated instrumentation in order to detect and quantify the minute amounts of genetic target available from a patient sample. We have developed a novel, rapid, and specific signal amplification method that utilizes free radical photopolymerization to transduce signal from binding of complementary biomolecules into a macroscopically detectable response, thus eliminating the reliance of expensive instrumentation in the detection method. This approach involves coupling photointiators to DNA hybridization sites instead of fluorophores or other traditional reporter molecules to allow for the surface-initiated polymerization of detectable monomer exclusively at hybridization sites on a biochip surface. High propagation rates of acrylate monomer and low termination rates of propagating chains in highly crosslinked monomer formulations allow for high amplification factors to be achieved, as thick polymer films may be generated from minute amounts of DNA hybrids on the surface. Such an amplification system could be made inexpensive and robust while still providing a highly detectable response that is not achievable with other established amplification methods. In this communication, two separate amplification systems will be presented. The first involves coupling water soluble, photoclevable initiators onto streptavidin proteins in a high initiator to protein ratio. Incubation of this product onto commercialized biochips containing surface bound, biotin labeled DNA molecules followed by monomer contact and exposure to UV radiation has allowed for sensitive and unaided visual detection of the biotinylated DNA sequences. In effort to further reduce the detection limits and provide milder assay conditions, a second amplification system that relies on coupling visible light photoinitiators to a biochip surface and that is capable of polymerizing monomers that fluoresce has also been developed. Such a system has been used to achieve the unaided visual detection of complementary nucleic acid hybridizations of the K-Ras biomarker, whose known codon 12, 13 and 61 mutations have been shown to play a crucial role in the onset of several types of cancers, including lung, pancreatic, and colorectal cancer.

187e Microfabricated Electrochemical And Biochemical Sensors For Toxic Gases

Ilwhan Oh[1], Chelsea N. Monty[2], Mark A. Shannon[3], and Rich I. Masel[2]. (1) Department of Chemical and Biomolecular Engineering, University of Illinois at Urbana-Champaign, 294 RAL, MC-712, 600 S. Mathews Ave., Urbana, IL 61801, (2) Chemical and Biomolecular Engineering, University of Illinois at Urbana-Champaign, 294 RAL, MC-712, 600 S. Mathews Ave., Urbana, IL 61801, (3) Mechanical Science and Engineering, University of Illinois at Urbana-Champaign, 294 RAL, MC-712, 600 S. Mathews Ave., Urbana, IL 61801

There are strong demands for chemical sensors for gaseous toxic chemicals such as organophosphate and explosives. One of the approaches for high-selectivity sensor is to utilize a specific chemical or biological interaction of the target molecule. Recently we showed that the electrochemical sensor based on oxime chemistry provides high selectivity toward organophosphate.

A novel type of gas sensor is fabricated by combining microfabrication techniques and electrochemical transducer. The microchannel sensor we built is composed of liquid/gas microchannels separated by a nanoporous membrane. When oxime chemistry is adapted into the microchannel sensor, it gives electrochemical potential response to trace vapor (10 ppb) of gas molecules within ~10 sec. Double microchannel design further reduces potential drift and simplifies the sensor setup.

187f Characterization Of Electrospun Nano-Fibers For Biosensor Applications

Manju Misra[1], Steven R. Nartker[1], Sean Daugherty[2], Per Askeland[1], Hazel-Ann Hosein[1], Lawrence T. Drzal[1], and **Paul Satoh**[2]. (1) Composite Materials and Structures Center and Chemical Engineering and Materials Science, Michigan State University, 2100 Engineering Building, East Lansing, MI 48824-1226, (2) Neogen Corporation, 620 Lesher Place, Lansing, MI 48912

Electrospinning is a unique way to produce novel polymer nanofibers with diameters typically in the range from 10 nm to 1500 nm. Nanofibers can be made from a variety of solutions or melts. Cellulose Nitrate nanofibers with controlled diameters of 350 nm to 1200 nm were prepared by altering the polymer concentration from 6 wt% to 10 wt%. The solvent used was a 70:30 mixture of ethanol: acetone. The nanofibers produced will be used as membrane materials in biosensor devices. Preliminary data shows that the electrospun cellulose nitrate fibers offer an improved surface area and can therefore bind more protein than their commercial counterparts. This study established the dominant factor for controlling fiber size; solution viscosity, and will continue by analyzing the effect of fiber size and alignment on biosensor performance. This research is financially supported by the Michigan Economic Development Corporation (MEDC), 21st Century Jobs Fund Project Number: 06-1-P1-0262 entitled “Fundamental Studies on Multifunctional Electrospun Nanofibers for Sensor Applications”.

246a Ocular Drug Delivery Using Microneedles

Jason Jiang[1], **Samirkumar R. Patel**[1], Harvinder S. Gill[2], Deepta Ghate[3], Bernard E. McCarey[3], Dayle H. Geroski[3], Henry F. Edelhauser[3], and Mark R. Prausnitz[4]. (1) Chemical & Biomolecular Engineering, Georgia Institute of Technology, Atlanta, GA 30332, (2) Biomedical Engineering, Georgia Institute of Technology, 315 Ferst Drive, Atlanta, GA 30332, (3) Dept. of Ophthalmology, Emory University, Atlanta, GA 30322, (4) Chemical and Biomolecular Engineering, Biomedical Engineering, Georgia Institute of Technology, 311 Ferst Drive, Atlanta, GA 30332-0100

Drug delivery to the eye can be challenging for many reasons. Barriers such as pre-corneal drug loss from tear drainage through the nasolacrimal duct, poor corneal permeability, or absorption of drug into vessels of the conjunctiva prevent many drugs from effectively reaching their targets. Often the target, like the retina, is in the posterior portion of the eye. Conventional methods such as topical application or systemic administration are not very effective in overcoming these barriers and intravitreal injections pose safety concerns, because of the invasive nature of the procedure. We propose here the use of microneedles, i.e. micron scale needles, to administer drugs in a minimally invasive way into ocular tissues such as the sclera and cornea. We assess the capabilities of hollow microneedles to infuse a drug solution into ocular tissues. We also asses the ability of solid micorneedles coated with a drug formulation to release their drug coating within ocular tissues.

Hollow microneedles were used to asses the ability of infusing a sulforhodamine solution, nanoparticle suspension and microparticle suspension into human cadaver sclera in an in-vitro experiment. The hollow micorneedles delivered tens of microliteres of sulforhodamine solution and nanoparticle suspension at depths of hundreds of microns into the sclera without penetrating the tissue. Micorparticle delivery was also accomplished at this scale using either collagenase or hyaluronidase to break up densely packed collagen or glycosaminoglycans fibers within the sclera.

Solid microneedles were coated with sodium fluorescein and inserted into rabbit cornea in an in-vivo experiment. The sodium fluorescein concentration was monitored in the anterior portion of the eye over a 24 hour period. Concentrations were achieved in the eye at 70 times greater using our solid microneedles as compared to the conventional route of topical delivery of the same dose. A similar experiment using micorneedles coated with pilocaprine, which constricts pupil size and is used in the treatment glaucoma, was performed and provided similar results. A safety exam was performed by a clinical ophthalmologist and showed the eye had no adverse events or inflammatory response to the microneedle procedure.

These results show for the first time that microneedles are capable of delivering drugs into ocular tissues. The results also show that micorneedles, in a minimally invasive manner, can deliver drugs to provide higher bioavailability than conventional methods, such as topical application. Microneedles have the potential to deliver formulations that can provide controlled delivery for the treatment of diseases of the eye.

246b Fabrication, Stabilization, and Short-Term Biocompatibility of Multiphasic Nanocolloids

Joerg Lahann, Mutsumi Yoshida, and Kyung-Ho Roh. University of Michigan, Dept. of Chemical Engineering, 3074 H.H. Dow Bldg., Ann Arbor, MI 48109-2136

Advances in nanotechnology, in particular the development of novel types of nanoparticles, will result in advanced tools for biomedical research and clinical practice. One exciting aspect of future nanomaterial research will be the possibility to combine therapy and imaging in multifunctional nanoparticle designs. In this context, anisotropic particles with subcellular dimensions may offer so far unattainable capabilities, because they could provide access to directional information with respect to nanoparticle-cell interactions.

Our approach toward biphasic nanocolloids differs fundamentally from the previously introduced methods in that it takes advantage of polymer solutions, specifically solutions of FDA approved polymers. Using electrified polymer jets to create an anisotropic materials distribution in nanoparticles, we have developed an approach towards biphasic nanocarriers that was named "electrohydrodynamic co-jetting" or "electrified co-jetting". Electrified jetting is a process to generate liquid jets by use of electrostatic forces.

We have recently shown that the electrified jetting process can produce water-stable polymer particles with two distinct phases (1-4). To address a critical hurdle towards the application of these biphasic nanocolloids as imaging probes, short-term biocompatibility was evaluated using model cell culture systems. Exposure of human endothelial cells and murine fibroblasts to biphasic nanocolloids did not affect cell proliferation as determined by a colorimetric proliferation assay. Moreover, double staining with annexin V and propidium iodide and subsequent flow cytometric analysis indicated high cell viability, although slightly decreased viability was observed at the highest dose tested (1 mg particles/ 106 seeded cells). Particle internalization as well as surface binding occurred simultaneously for both cell types, as evidenced by confocal laser scanning microscopy. Taken together, these results suggest excellent short-term biocompatibility in physiological systems for wide concentration ranges of the biphasic nanocolloids and open possibilities for future work investigating receptor- or surface marker-mediated targeting.

[1] K. H. Roh, D. C. Martin, J. Lahann, Nat. Mater. 2005, 4, 759. [2] K. H. Roh, M. Yoshida, J. Lahann, Langmuir 2007, in press. [3] M. Yoshida, K. H. Roh, J. Lahann, Biomaterials 2007, in press. [4] K. H. Roh, D. C. Martin, J. Lahann, J. Am. Chem. Soc. 2006, 128, 6796.

246c Surface Functionality Affects Transport Mechanisms of Dendrimer-Based Drug Delivery Nanodevices

Rajyalakshmi Inapagolla, Wayne State University (Department of Chemical Engineering), 5050 Anthony Wayne Drive, Detroit, MI 48202, Omathanu P. Perumal, College of Pharmacy, South Dakota State University, Brookings, SD 57007, Sujatha Kannan, Pediatrics, Children's Hospital of Michigan, Wayne State University (Department of Chemical Engineering), Detroit, MI 48202, and **Rangaramanujam M. Kannan**, Wayne State University, 1121 Engineering, 5050 Anthony Wayne Drive, Detroit, MI 48202.

Purpose: Dendrimers are an emerging class of nanoscale drug delivery vehicles, with a potential to delivery therapeutics in a targeted manner. They present a large density of functional groups at the surface that play a key role in the cellular uptake and eventual drug release. The cellular uptake, and the eventual The aim of this study is to understand the influence of surface charge on cellular interactions and intracellular trafficking of polyamidoamine (PAMAM) dendrimers in lung epithelial cells.

Methods PAMAM dendrimers with cationic (G4 amine terminated), anionic (G3.5 carboxyl terminated) and neutral (G4 hydroxyl terminated) surface charge were labeled with fluoroisothiocyanate (FITC). Cell uptake and subsequent intracellular fate was determined by flow cytometry, fluorescence and confocal microscopy using A549 lung epithelial cells. Temperature block and metabolic inhibition (sodium azide and 2-deoxyglucose) were used to investigate if the cellular uptake was an energy dependent mechanism. Various endocytotic blockers including chlorpromazine, sucrose, filipin, nocodazole, ammonium chloride and cytochalasin were used to identify the pathways involved in the transport of different dendrimers.

Results Cellular uptake of all dendrimers increased as a function of incubation time, but the uptake of cationic and neutral dendrimers was much higher than anionic dendrimers. Incubation at 100C and metabolic inhibition significantly reduced the cellular uptake of dendrimers. All dendrimers were found to be localized in lysosomes within 30 minutes. Sucrose reduced the cellular uptake of all the dendrimers by more than 50%. Chlorpromazine, filipin, nocodazole and ammonium chloride did not show any inhibitory effect on endocytotic uptake of cationic and neutral dendrimers, but filipin, noocodazole, and ammonium chloride significantly reduced the cellular uptake of anionic dendrimer.

Conclusions The surface charge and functionality play a significant role on the cellular interactions and subsequent endocytotic uptake into the cell. Anionic dendrimers appear to be mainly taken up by caveolae mediated endocytosis in A549 lung epithelial cells. While cationic and neutral dendrimers are taken in by a non-clathrin and non-caveolae endocytosis mechansim and may be they are transported by electrostatic or non specific fluid phase endocytosis.These findings open up new possibilities of targeting therapeutic agents to specific cell organelles based on surface charge.

Web Page: www.kannangroup.com

246d Integrin Alpha5Beta1 Targeted Delivery and Controlled Release to Colon Cancer Cells

Ashish Garg, Alison Wedekind, and Efrosini Kokkoli. Department of Chemical Engineering and Materials Science, University of Minnesota, Minneapolis, MN 55455

Integrin alpha(5)beta(1) is expressed on several types of cancer cells, including colon cancer, and play an important role in tumor growth and metastasis. The ability to target the integrin alpha(5)beta(1) using an appropriate drug delivery nanovector can significantly help in inhibiting tumor growth, reducing tumor metastasis, and decreasing deleterious side effects associated with different cancer therapies. Liposomes are nanosized phospholipid bilayer vesicles that have been extensively studied as drug delivery carriers. The goal of this study is to design peptide functionalized stealth liposomes (liposomes covered with polyethylene glycol (PEG)) that will specifically target the integrin alpha(5)beta(1) and effectively release their encapsulated load intracellularly. The PEG provides a steric barrier allowing the liposomes to circulate in the blood and the functionalizing moiety, PR_b peptide, will specifically recognize and bind to alpha(5)beta(1) expressing cells. PR_b is a novel peptide sequence that was designed in our lab and mimics the cell adhesion domain of fibronectin. It includes both RGDSP (the primary recognition site for alpha(5)beta(1)) and PHSRN (the synergy site for alpha(5)beta(1)) connected with a (SG)5 linker. In this study we have demonstrated that by optimizing the amount of PEG and PR_b on the liposomal interface we can engineer nanovectors that bind to mouse and human colon cancer cells in a specific manner and are internalized through alpha(5)beta(1)-mediated endocytosis. GRGDSP-targeted stealth liposomes bind to colon cancer cells and internalize, but they have much lesser efficiency than PR_b targeted stealth liposomes. We have further shown that these liposomes upon internalization are capable of releasing their load intracellularly. The proposed targeted delivery system has a great potential to deliver therapeutic agents directly to colon cancer cells, in an efficient and specific manner.

246e 3D Self-Assembled Microdevices for Cell Therapy and Chemical Delivery

Timothy Leong[1], Christina Randall[2], Noy Bassik[1], and **David H. Gracias**[1]. (1) Department of Chemical and Biomolecular Engineering, Johns Hopkins University, 125 Maryland Hall, 3400 N Charles Street, Baltimore, MD 21218, (2) Department of Biomedical Engineering, Johns Hopkins University, Baltimore, MD 21218

We describe a new class of biomedical microdevices based on self-assembling 3D polyhedra. The polyhedra are fabricated from lithographically structured 2D templates and fold on a thermal trigger. We have utilized the polyhedra as containers for on-demand remote controlled delivery to live cells and in tissue. We will also describe experiments in cell encapsulation and imaging.

Web Page: www.jhu.edu/chbe/gracias/

253a Nanoparticle-Modulated Micro-Retroreflectors

Steve Kemper[1], Tim Sherlock[2], Paul Ruchhoeft[2], and **Richard C. Willson**[3]. (1) Chemical Engineering, University of Houston, CHEE 4004 4800 Calhoun Rd., Houston, TX 77004-4004, (2) Electrical and Computer Engineering, University of Houston, CHEE 4004 4800 Calhoun Rd., Houston, TX 77004-4004, (3) University of Houston, CHEE 4004 4800 Calhoun Rd., Houston, TX 77004-4004

We have fabricated micron-sized retroreflectors (structures that return incident light directly back to the source) and have shown that they are extremely bright and detectable over a large range of angles when inspected with a simple optical microscope. These retroreflectors are part of an ultra-sensitive detector platform for sensing small quantities of virus particles, bacteria, DNA, RNA, or any variety of molecules of interest. In particular, we are targeting our first generation sensor to detect Norwalk virus particles. In this proposed system, the base of the retroreflector is decorated with antibodies to the virus, and, if present, the virus particles are captured by this surface. After capture, gold nanoparticles, coated with a secondary antibody, are introduced into the system, attach to the virus, and drastically reduce the retroreflector brightness with a specific, well-understood spectral signature. If no virus is present, the reflectivity is unaffected. We have measured the reflectivity versus 40nm diameter gold nanoparticle surface density and have found a 40% reduction in signal for 100 nanoparticles per square micron when illuminated with broad-band light. The base of the retroreflector is about 4 square microns in size, yielding a detection limit of hundreds of particles. Further optimization of particle size and illumination wavelength is expected to increase this sensitivity substantially. Retroreflectors are fabricated by coating a silicon wafer with 2.5 microns of polyimide and 200nm of resist. A lithography step is used to generate the retroreflector pattern as openings in the resist and a 50nm thick nickel coating is deposited using thermal evaporation. After a lift-off step, which leaves behind only the nickel that coated the base of the resist openings, the patterns are transferred into the polyimide in an O2/CF4 reactive ion etch, leaving the retroreflecting structure with very straight relatively smooth walls. Next, gold is evaporated to coat the base of the structure and a directional evaporation step is used to cover all but the sensor base with aluminum. The gold is selectively functionalized with amine-reactive thiol molecules which serve as a platform for attaching antibodies, oligonucleotides, or other detector molecules.

253b Nanoparticle-Based Assay For DNA Detection And Quantification With Single Nucleotide Discrimination Selectivity

Weijie Qin and **Lin Yue Lanry Yung**. Department of Chemical and Biomolecular Engineering, National University of Singapore, Singapore, Singapore

Quantification and selectivity are the other two important aspects , besides, sensitivity, for the evaluation of DNA biosensor devices. DNA quantification is critical for gene expression analysis, detection of DNA mutations or genetic defects, early stage diagnosis of critical illness such as HIV and cancers, and forensic applications. Furthermore, diagnosis of pathogenic and genetic diseases requires the device to have high selectivity that can discriminate single nucleotide mismatches. Single nucleotide polymorphisms (SNPs) are the most abundant form of genetic variation that occur once every 100–300 bases and there are greater than 3 million SNPs in the human genome. Identify these SNPs and associate individual SNPs with specific diseases and pharmacological responses are clinically important for medical diagnostics, disease prevention and prognostics. Currently, real-time polymerase chain reaction (RT-PCR) is one of the most frequently used methods for DNA quantification and SNP discrimination in life science and clinical research. However RT-PCR is a time-consuming and labor intense process, and its selectivity is not always satisfactory even with sophisticated optimizations. Previously, we have shown that gold nanoparticle-DNA (nAu-DNA) conjugates bearing definite number of short DNA (< 20 bases) can be prepared by gel electrophoresis isolation followed by restriction endonuclease manipulation of the nAu-bound DNA. In this study, we described a novel gold nanoparticle (nAu) based assay methodology that has reliable quantification ability and SNP discrimination selectivity. In this assay, two sets of specially designed nAu-DNA conjugates are fabricated via the gel electrophoresis and restriction endonuclease manipulation methods. These two sets of conjugates with definite number (1, 2, 3…) of short single stranded DNA (ssDNA) probes are not directly complementary to each other. After mixing, these conjugates do not recognize and group each other until a target DNA that is complementary to both sets of conjugates is introduced. Only conjugate groupings with defined structure (dimers or trimers) can form due to definite number of DNA strands on each nAu. The resulting conjugate groupings are characterized and quantified by agarose gel electrophoresis. The size differentiation ability of gel electrophoresis allows strict discrimination between different conjugate structures (monomer, dimer and trimer) and enables precise quantification of target DNA samples. Furthermore, a strong discrimination between perfectly matched and single base mismatched DNA is achieved since only the perfectly matched target DNA allows the formation of conjugate groupings. This novel assay methodology is particularly suitable for quantitative DNA analysis and SNP discrimination, and is expected to find application in the diagnosis of genetic & infective disease, forensic analysis as well as biodefense applications.

253c Surfaces: Co-Localization Of Oligonucleotides And Gold Nanoparticles, A Novel Approach Enabling Pcr

Udayarka Karra, CHEMICAL ENGINEERING, UNIVERSITY OF UTAH, 50 SOUTH CENTRAL CAMPUS DRIVE,, SALT LAKE CITY, UT 84112

Gold nanoparticles (AuNPs) have unique optical and energetic properties enabling their use in analytic and biosensing applications through Surface Plasmon Resonance (SPR) and Polaritons. Oligonucleotides are functionalized on the 5' and 3' end with alkylthiol or any thiol activated compounds for their high affinity towards AuNPs. The oligos are used in conjugation with stabilized AuNPs as a mode to detect their derivitization on mercaptosilanized quartz surfaces, which as far has been the most effective and active surface. The other method to derivitize oligos is modifying them to directly couple with active gold surfaces without any aid of substrate or functionality. The objective of the study is to apply the above methods for their successful imitation to optical fibers, which under the controlled conditions of a PCR and the thermal excitation of AuNPs for the essential stable polymerization of oligonucleotides without denaturing them. The work on the immobilization of oligonucleotides on plain/gold activated surfaces, and their detection by the aid of AuNPs, is under investigation and refinement for application to optical fibers.

253d Nanobiosensor Utilizing Genetically Engineered Receptor Proteins Immobilized Within Au Nanopores

Ian I. Suni[1], Abhinav Tripathi[2], Jianbin Wang[2], and Linda A. Luck[3]. (1) Department of Chemical and Biomolecular Engineering, Clarkson University-, Potsdam, NY 13699-5705, (2) Chemical and Biomolecular Engineering, Clarkson University-, Potsdam, NY 13699-5705, (3) Department of Chemistry, State University of New York, 319 Hudson Hall, 101 Broad Street, Plattsburgh, NY 12901

A new type of nanopore sensor design is reported for a reagent-less electrochemical biosensor with no analyte "tagging" by fluorescent molecules, nanoparticles, or other species. This sensor design involves immobilization within Au-coated polycarbonate nanopores of bacterial periplasmic binding proteins (bPBP), which undergo a wide amplitude hinge-twist motion upon ligand binding. Ligand binding thus triggers a reduction in the effective thickness of the immobilized protein film, which is detected as an increase in electrolyte conductivity (decrease in impedance) through the nanopores.

This new sensor design is demonstrated for glucose detection using a cysteine-tagged mutant (GGR Q26C) of the galactose/glucose receptor (GGR) protein from the bPBP family. The GGR Q26C protein is immobilized onto Au nano-islands that are deposited within the pores of commercially available nanoporous polycarbonate membranes.

Results are presented for glucose concentrations ranging from 40-200 ìM and for pore diameters of 10, 30 and 100 nm. As expected, the sensitivity to glucose is reduced at concentrations much greater than the equilibrium ligand binding constant, about 0.2 ìM for the soluble form. The binding constant can be increased into the physiological range through residue substitution near the ligand binding pocket. The greatest sensitivity to glucose is found for the smallest pore diameter (10 nm).

253e Spectroscopic Characteristization Of Plasmon Capillaries Coated With Gold By Electroless Plating

Wonmi Ahn and D. Keith Roper. Chemical Engineering, University of Utah, 50 So. Central Campus Dr., 3290 MEB, Salt Lake City, UT 84112

Optical light guiding capillary is used for evanescent wave absorption spectroscopy due to its unique geometry, consisting of hollow center surrounded by fused-silica capillary core, doped-silica cladding, and polyimide protection coating, respectively. Capillary guides light in the wavelength region from UV to near infrared through the wall of the capillary allowing long pathlength, small sample requirement, and direct contact of sample with a sensing area. Here, we report fabrication of a plasmon capillary by plating gold on inner wall of the optical capillary using an electroless gold plating method. Uniform gold deposition occurs over the entire inner core surface where propagating light induces surface plasmon resonance (SPR). Effects of silica core size, and plated gold thickness are examined to maximize the SPR response. Amplitude, breadth and red-shift of absorbance spectra are characterized for liquid and gas samples. Fundamental data in this study are applied to applications in chemical sensing and optical microfluidic heating.

253f Gold Nanoparticle Amplified DNA Hybridization Detection by Electrochemical Impedance Spectroscopy (Eis)

Hong Susan Zhou, Chemical Engineering, Worcester Polytechnic Institute, 100 Institute Road, Worcester, MA 01609

Rapid development of microanalytical devices has provided convenience and capability for disease diagnosis, drug screening, or forensic applications. Optical detection of DNA hybridization usually needs to detect a fluorescent signal by large and expensive optical equipment. Electrochemical detection offers great promise for rapid, simple and low-cost detection of specific nucleic acid sequences. A label such as enzyme labels or electroactive labels have been introduced to improve the sensitivity of the electrical assays. However, development of methods that do not require additional instrumentation and reagents is still a challenge for point-of-care or field use. Efforts on label-free electrical DNA hybridization have encountered problems of random orientation of thiol-modified oligonucleotide probes thus unreliable signal output. Here we report a nanoparticle-based electrical detection of DNA hybridization. A well-controlled mixed self-assembled monolayer method is adopted for the immobilization of thiolated-ssDNA. Electrochemical Impedance Spectroscopy is applied for the detection of DNA hybridization on gold electrodes. By using AFM, we proved that Au nanoparticles are on the biosensor surface for specific DNA hybridization after stringency wash. The results indicate that nanoparticle-based electrical DNA hybridization detection provides a rapid, robust, low-cost and sensitive detection scheme which has great potential for hand-held microarray technologies.

366a Enzymes Supported in Mesoporous Molecular Sieve Fibers

Kenneth J. Balkus Jr., Minedys Macias, Daniel Tram, and Harvey Lui. Chemistry, University of Texas at Dallas, P. O. Box 830688, Richardson, TX 75083

The fixation of biologically active species onto inorganic materials would combine the high selectivity of enzymatic reactions with the chemical and mechanical properties of the support. This blend has brought to light many new applications in the fields of biofuel cells, chemical sensing and biocatalysis. Molecular sieves as support materials offer interesting properties, such as high surface areas, hydrophobic or hydrophilic behavior and electrostatic interactions, as well as, mechanical and chemical resistance which makes them attractive for enzyme immobilization. Emerging applications have generated a greater demand for new ways to manipulate and configure these nanoporous materials. The required form of a molecular sieve may involve dramatically different length scales, ranging from shaped particles to fibers and continuous coatings. The formation of free standing mesoporous molecular sieve fibers as spider webs and papers has been achieved by electrostatic deposition. Enzymes have been immobilized by incorporating the biomolecule into the sol gel precursor. Alternatively, a core shell fiber can be prepared where the enzyme is encapsulated in the core. Recent results involving glucose oxidase and cellulase as well as microperoxidase will be described.

366b Crosslinked Enzyme Aggregates in Hierarchically-Ordered Mesoporous Silica: A Simple and Effective Method for Enzyme Stabilization

Moon Il Kim[1], Jungbae Kim[2], Ho Nam Chang[3], and **Hyun Gyu Park**[3]. (1) Department of Chemical and Biological Engineering, Rensselaer Polytechnic Institute, Center for Biotechnology and Interdisciplinary Studies, Room 4211, 110 8th Street, Troy, NY 12180-3590, (2) Pacific Northwest National Laboratory, 902 Battelle Blvd., PO Box 999, Mailstop K8-93, Richland, WA 99352, (3) Department of Chemical and Biomolecular Engineering, Korea Advanced Institute of Science & Technology, 373-1 Guseong-dong, Yuseong-gu, Daejeon, 305-701, South Korea

α-chymotrypsin (CT) and lipase (LP) were immobilized in hierarchically-ordered mesocellular mesoporous silica (HMMS) in a simple but effective way for the enzyme stabilization, which was achieved by the enzyme adsorption followed by glutaraldehyde (GA) crosslinking. This resulted in the formation of nanometer scale crosslinked enzyme aggregates (CLEAs) entrapped in the mesocellular pores of HMMS (37 nm), which did not leach out of HMMS through narrow mesoporous channels (13 nm). CLEA of α-chymotrypsin (CLEA-CT) in HMMS showed a high enzyme loading capacity and significantly increased enzyme stability. No activity decrease of CLEA-CT was observed for two weeks under even rigorously shaking condition, while adsorbed CT in HMMS and free CT showed a rapid inactivation due to the enzyme leaching and presumably autolysis, respectively. With the CLEA-CT in HMMS, however, there was no tryptic digestion observed suggesting that the CLEA-CT is not susceptible to autolysis. Moreover, CLEA of lipase (CLEA-LP) in HMMS retained 30% specific activity of free lipase with greatly enhanced stability. This work demonstrates that HMMS can be efficiently employed as host materials for enzyme immobilization leading to highly enhanced stability of the immobilized enzymes with high enzyme loading and activity.

Immobilization of Lipase on Hydrophobically Modified Siliceous Mesocellular Foam Under High Flow Condition

Yu Han, Su Seong Lee, and Jackie Y. Ying*

Institute of Bioengineering and Nanotechnology, 31 Biopolis Way, The Nanos, Singapore 138669

Enzyme-catalyzed asymmetric reactions have attracted a lot of interest as one of the most important fields in organic and pharmaceutical syntheses.[1-7] Immobilization of enzymes onto solid supports may render these catalysts more mechanically robust and thermally stable, and allow for ease of separation from the reaction media.[2-5] Through immobilization, the catalytic activity of these systems may also be enhanced due to reduced enzyme aggregation, especially in non-polar organic media.[6,7]

MCF and FDU-12 were chosen as the support materials for enzyme immobilization study due to several reasons. First, they have large and tunable pore sizes (20–50 nm and 10–15 nm, respectively), which can accommodate enzymes of different dimensions.[8] Secondly, MCF and FDU-12 both possess a 3-dimensional, interconnected pore structure, which would facilitate mass transfer and molecular diffusion. Thirdly, their pores are cage-like, and interconnected by windows smaller than the diameter of the cage.[9,10] This would allow for the entrapment of enzymes with sizes close to (or slightly larger than) that of the windows (Scheme 1), while minimizing enzyme leaching from the pores due to the restriction of the small windows.

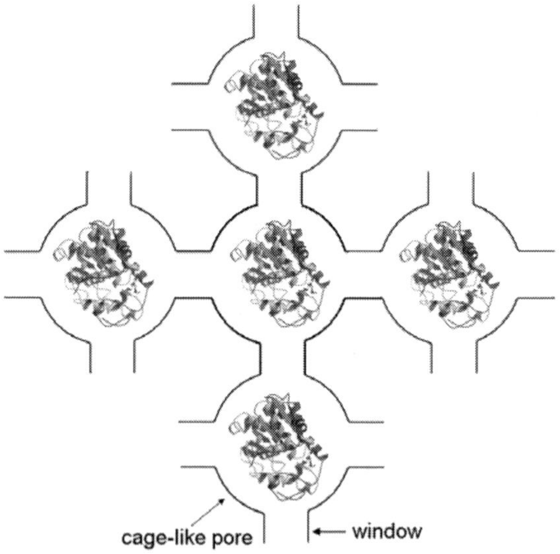

Scheme 1. Enzymes entrapped in the cage-like mesopores of MCF and FDU-12. (This scheme gives an ideal representation of the entrapment of enzymes in the cage-like pores. The hydrophobic interaction between the enzymes and the long-chain alkyl groups (C_8 or C_{18}) on silica surface has been omitted.)

Considering the dimensions of CALB (6.9 nm × 5.0 nm × 8.7 nm), MCF and FDU-12 with window sizes of ~ 10 nm were prepared by controlling the synthesis conditions. Conventional MCF has an irregular particle morphology with a very broad particle size distribution.[9] By modifying the synthesis conditions,[11] we have derived spherical MCF particles with a fairly uniform particle size of ~ 5 μm (Figure 1a). Figure 1b illustrates the high porosity and uniform pore size of the spherical MCF particle. The MCF sample was ~ 25 nm in pore diameter (Figure 1c), which was consistent with the average pore size (26.7 nm) obtained from N_2 adsorption isotherm. The window size was determined from the N_2 desorption isotherm to be 14.5 nm by the BdB-FHH method.[12] The MCF particles possessed a pore volume of 2.16 cm^3/g and a BET surface area of 535 m^2/g (Table 1).

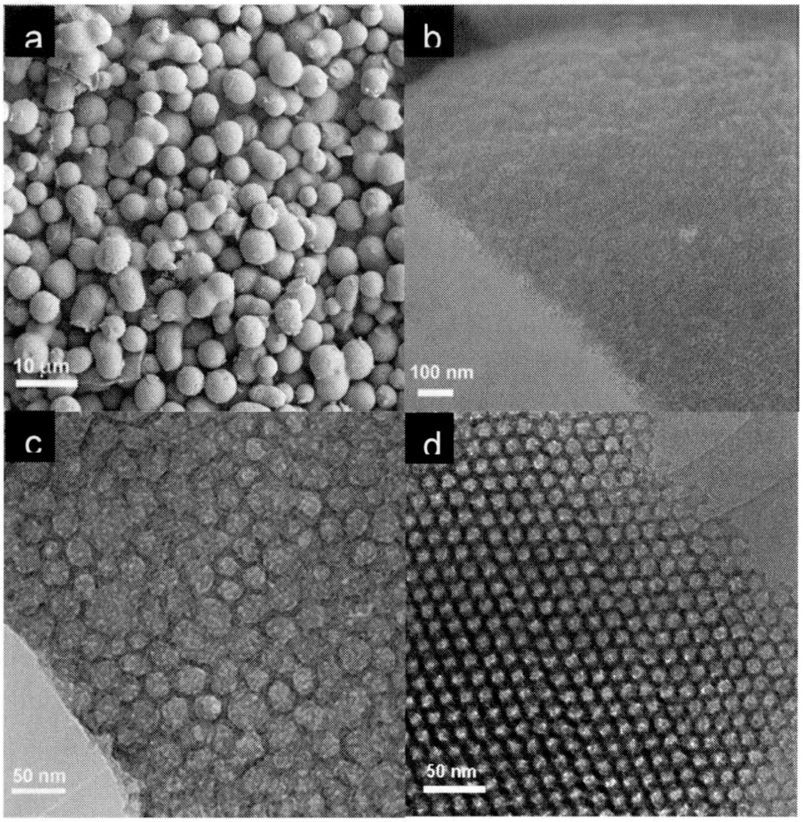

Figure 1. SEM micrographs of (a) spherical MCF particles, and (b) a single spherical MCF particle at high magnification. TEM micrographs of (c) MCF, and (d) FDU-12 taken at [110] incidence.

The modification of MCF with C_8 and C_{18} groups decreased both the pore diameter and the window diameter by ~ 2 nm and ~ 4 nm, respectively (Table 1). Although C_8-MCF and C_{18}-MCF showed lower surface area and pore volume than the unmodified MCF (Table 1), they were still highly porous with narrow pore size distributions. FDU-12 has a pore size and a window size of 14 nm and 9.0 nm, respectively, (Figure 1d, Table 1). Its pore volume and BET surface area were 1.25 cm^3/g and 509 m^2/g, respectively. FDU-12 was modified only by C_8, which decreased the window size to 7.5 nm. The window size would be too small to entrap CALB if FDU-12 were to be modified by C_{18}.

Compared to the conventional approach, the pressure-driven method resulted in significantly higher enzyme loading on C_8-MCF, C_{18}-MCF and C_8-FDU-12 within a shorter period of time (Figure 2). For example, 24 h of stirring led to an enzyme loading of 92 mg/g onto C_8-MCF, whereas an enzyme loading of 275 mg/g onto C_8-MCF was achieved in 2 h with the pressure-driven method. C_8-FDU-12, which has smaller pore size and pore volume than C_8-MCF only provided about half the enzyme loading as C_8-MCF.

Table 1. Characteristics of the mesoporous silica materials before and after modification and loading with CALB by the pressure driven method.[a]

	Pore Size (nm)	Window Size (nm)	Surface Area (m^2/g)	Pore Volume (cm^3/g)
MCF	26.7	14.5	535	2.16
C_8-MCF	24.3	12.5	332	1.60
C_{18}-MCF	22.5	10.6	252	1.27
CALB/C_{18}-MCF	20.0	9.3	208	1.05
FDU-12	14.0	9.0	509	1.25
C_8-FDU-12	12.0	7.5	319	0.81
CALB/C_8-FDU-12	12.0	6.0	194	0.50

[a]Determined from N_2 sorption isotherms. The pore sizes were calculated from the adsorption branches based on the BdB sphere model. The window sizes were calculated from the desorption branches based on the BdB sphere model.

However, C_8-FDU-12 also demonstrated much higher enzyme loading by the pressure-driven method than by the conventional method (Figure 2). When the concentration of CALB stock solution was reduced from 8 to 4 mg/ml, the resulting enzyme loading was slightly lowered in all cases. Nevertheless, the pressure-driven method still loaded significantly more enzymes onto the three different hydrophobic mesoporous supports than the conventional stirring approach. Thus, while the enzyme loading was affected by the concentration of enzyme stock solution, it was most significantly impacted by the method of enzyme loading. Using the pressure-driven method, substantially more enzymes could be loaded onto the mesoporous silica supports, especially those that have ultralarge pore sizes and high pore volumes. The aqueous enzyme stock solution possibly did not have a high affinity with the hydrophobic supports, thus, it was more effective to force the enzymes to enter the mesopores by high pressures.

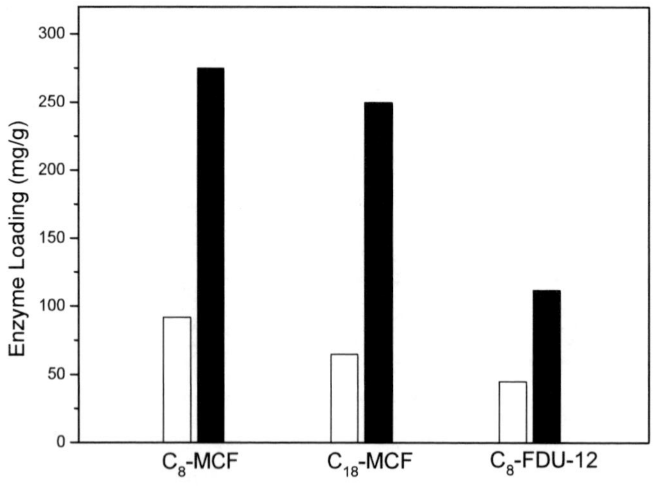

Figure 2. CALB loading on the modified mesoporous silica supports using (□) the conventional stirring method and (■) the pressure-driven method.

CALB/C_{18}-MCF prepared by the pressure-driven method still exhibited high BET surface area (208 m^2/g) and pore volume (1.05 cm^3/g) (Table 1). The pore diameter of C_{18}-MCF was only reduced from 22.5 nm to 20.0 nm after the enzyme loading, possibly because the tertiary structure of CALB was unfolded under the vacuum treatment associated with the N_2 adsorption-desorption experiment.[4c] CALB/C_{18}-MCF also showed PA-FTIR peaks at 1650 cm^{-1} and 3300 cm^{-1}, which were associated with carbonyl and amino groups, respectively, confirming the enzyme incorporation. The surface area and pore volume of C_8-FDU-12 were decreased significantly by ~ 40% after CALB immobilization (Table 1), suggesting that some of the pores in this mesoporous support were blocked by the enzymes due to the small window size of C_8-FDU-12.

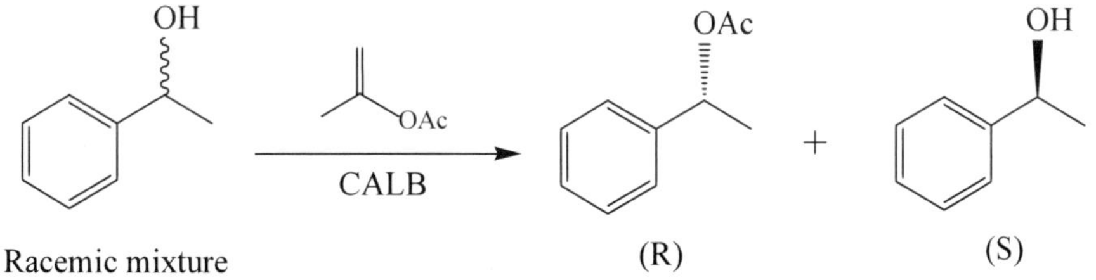

Scheme 2. Kinetic resolution of 1-phenylethanol with isopropenyl acetate by CALB.

For the acylation of 1-phenylethanol by isopropenyl acetate (Scheme 2), a fixed amount of CALB was introduced as free or supported catalysts. CALB/C_8-MCF and CALB/C_{18}-MCF prepared by the pressure-driven method showed similar catalytic activity (Figure 3). In both cases, complete conversion of (R)-1-phenylethanol to (R)-1-phenylethyl acetate (i.e. 50% conversion of the (R)-1-phenylethanol and (S)-1-phenylethanol racemic mixture) was achieved after 5 h. In contrast, free CALB showed a slower reaction rate, giving 45.5% conversion in 5 h. This is possibly because the free CALB formed aggregates in toluene, leading to a lower active surface area. By loading CALB uniformly within the pores of MCF, aggregation was avoided or reduced. The weak hydrophobic interaction between the immobilized CALB and the pore surface of MCF allowed CALB to maintain its active conformation and therefore high activity. CALB/C_8-FDU-12 gave an even lower reaction rate (40% conversion in 5 h) than free CALB (Figure 3). This could be due to the relatively low porosity and the blocked porous network of C_8-FDU-12 after CALB immobilization (Table 1), limiting the mass transfer in the catalytic reaction. This study indicated that MCF was more suitable than FDU-12 for CALB immobilization given its high porosity and ultralarge pore size. CALB/C_8-MCF, CALB/C_{18}-MCF and CALB/C_8-FDU-12 prepared by the pressure-driven method all showed similarly high enantioselectivities ($ee_s > 99.7\%$ and $ee_p > 99.7\%$) as the free CALB. This confirmed that the pressure-driven method for enzyme loading did not alter the selectivity of CALB.

CALB/C_{18}-MCF prepared by the conventional stirring method showed a substantial decrease in 6-h conversions (from 50% to 37%) over 5 runs (Figure 4a). Elemental analysis revealed that over 65% of the CALB loaded in C_{18}-MCF was lost after 5 runs, suggesting that the decrease in activity was mainly due to enzyme leaching. In contrast, CALB/C_{18}-MCF prepared by the pressure-driven method only showed a minor decrease in conversion (from 50% to 46.5%) after 8 runs or 80 h of reaction (Figure 4b). Elemental analysis showed that only 10% of the CALB loaded by pressure has leached out of the C_{18}-MCF support after 8 runs. Thus, the pressure-driven method not only significantly improved the enzyme loading, but also greatly reduced the enzyme leaching.

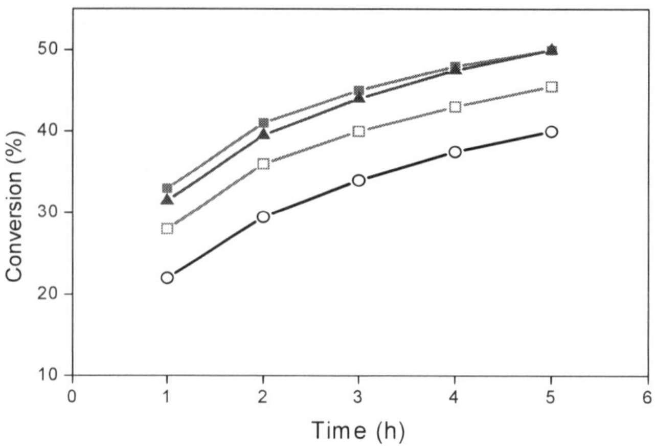

Figure 3. Catalytic conversions over (□) free CALB, and (▲) $CALB/C_8$-MCF, (■) $CALB/C_{18}$-MCF and (○) $CALB/C_8$-FDU-12 prepared by the pressure-driven method.

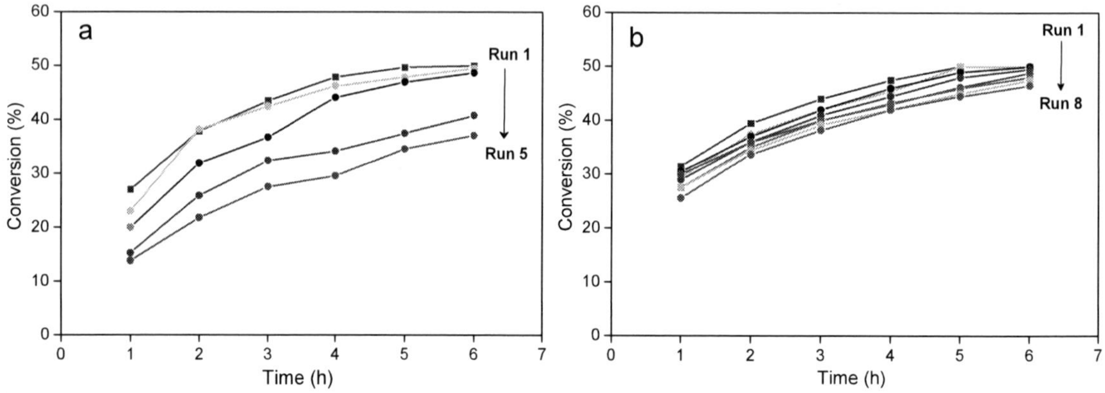

Figure 4. Conversion vs. time in sequential runs over $CALB/C_{18}$-MCF prepared by (a) the conventional immobilization method and (b) the pressure-driven method.

The enzyme molecules were driven by pressure to settle deep within the porous framework of MCF. The cage-like pores of MCF physically inhibited the entrapped CALB from leaching. In contrast, a substantial amount of the CALB molecules was probably weakly adsorbed on the outer surface of the mesoporous particles when the conventional stirring method was used. These enzymes would be leached from the support relatively easily during the reaction runs.

In a separate experiment, we immobilized CALB onto C_{18}-MCF using the pressure-driven method in a HPLC column, and then flushed the as-prepared column with 500 ml of PBS buffer solution and then 1000 ml of deionized water under a moderate pressure of ~ 1000 psi. Elemental analysis indicated that such a thorough washing hardly decreased the CALB loading in C_{18}-MCF; the PA-FTIR C=O stretching peak was also not affected in intensity. This finding confirmed that the enzymes were firmly entrapped within the modified MCF support by the pressure-driven method, and were robust against leaching.

The pressure-driven method also resulted in CALB/C_{18}-MCF with a higher catalytic activity than the stirring method. For example, in the first run, the catalysts prepared by pressure-driven method and stirring method showed 1-h conversions of 31.5% and 27%, respectively, although the same amount of CALB was introduced to both reaction systems (Figure 4). This could be because a more homogeneous dispersion of CALB was achieved throughout the C_{18}-MCF particles by the pressure-driven method, whereas some agglomeration or multilayer adsorption of CALB might be present at the outer shell of the support particles with the stirring method.

The reusability of CALB/C_8-MCF and CALB/C_{18}-MCF prepared by the pressure-driven method was compared to Novozyme 435. Since the actual enzyme loading in Novozyme 435 was unknown, we adjusted the amounts of catalysts introduced to the reaction system so that all three catalysts gave 33% conversion after 1 h in the first run. Compared to Novozyme 435, CALB/C_8-MCF and CALB/C_{18}-MCF showed superior 1-h conversions over subsequent runs, illustrating that CALB immobilized onto mesoporous silica by the pressure-driven method has better long-term stability and reusability than the commercial CALB/polymer catalyst.

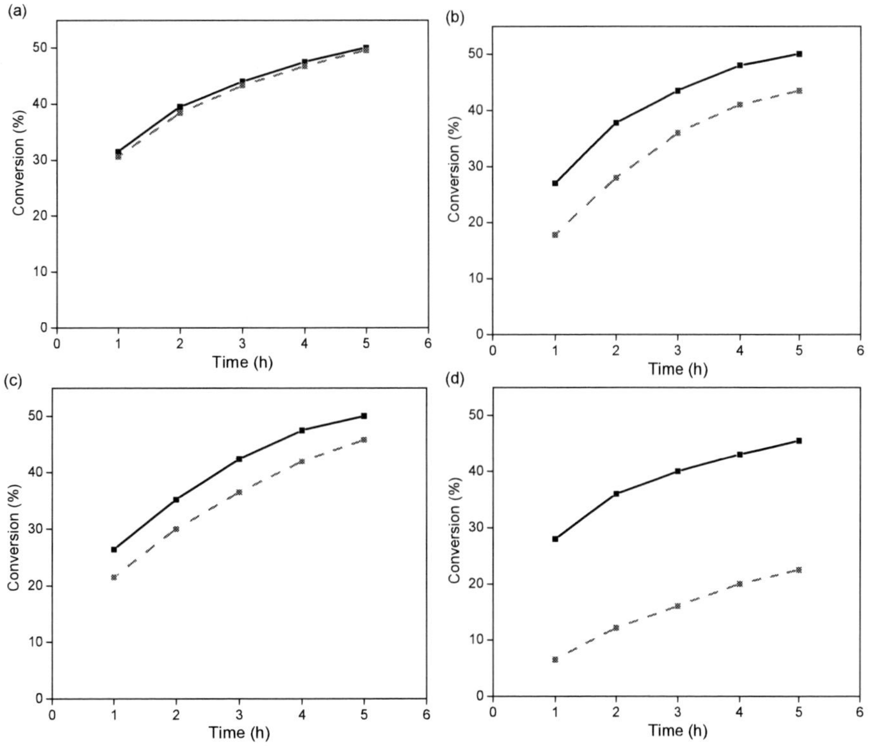

Figure 5. Conversion vs. time over CALB/C_{18}-MCF prepared by (a) the pressure-driven method and (b) the conventional method, (c) Novozym 435, and (d) free CALB. Catalysts were tested (—) before and (---) after heat treatment at 80°C.

The catalytic applications of enzymes are often hampered by their poor thermal stability, which may be improved through enzyme immobilization. To examine their thermal stability, free and supported CALB catalysts were tested for catalytic activity before and after 15 h of heat treatment at 80°C. CALB/C_{18}-MCF prepared by the pressure-driven method showed negligible loss in activity from the heat treatment (Figure 5(a)). In contrast, CALB/MCF-C_{18} prepared by the conventional stirring method and Novozyme 435 showed a substantial reduction in activity after the thermal treatment

(Figures 5(b) and 5(c)). Free CALB was severely deactivated upon heat treatment and lost most of its activity (Figure 5(d)). The deactivation of CALB at high temperature was attributed to the change of its conformation (enzyme distortion).[4c] Since the pore size of the MCF support was similar to the dimensions of CALB, the entrapped CALB was confined within the mesopores, which might suppress the change in enzyme conformation at high temperature. When immobilized by the conventional stirring method, some of the CALB would be located on the outer surface of the mesoporous silica particles; these enzymes would not be properly stabilized against heat treatment. This study demonstrated the effectiveness of the pressure-driven method in enzyme immobilization to achieve enhanced thermal stability for enzyme catalysts.

A standard HPLC column (250 mm × 4.6 mm) was employed in the preparation of CALB/C_{18}-MCF by the pressure-driven method. The catalyst-packed column was then tested for continuous reaction. Full conversion of (R)-1-phenylethanol to (R)-1-phenylethyl acetate was achieved even at a fairly high flow rate of 1.5 ml/min, giving a high initial reaction rate of 3.2 mM 1-phenylethanol/(min·g enzyme). The activity and enantioselectivity of the reaction remained constant during the first 10 h of reaction. After 48 h of reaction at a constant flow rate of 1.5 ml/min, the reaction rate decreased by 9%. It should be noted that our spherical MCF particles were particularly suitable for use in a packed bed reactor since their uniform size and morphology allowed for homogeneous packing with a moderate backpressure. For example, at a flow rate of 1.5 ml/min, the pressure drop over a 25 cm-long column was 1050 psi. Commercial Novozyme 435 was also not suitable for packed bed reactor since its polymer support would swell in the organic solvents, resulting in an unstable pressure in the reactor.

This study illustrated that spherical MCF particles served as an excellent support for enzyme immobilization. The ultralarge cage-like mesopores of this silica support were ideal for entrapping enzymes. A pressure-driven method was developed to effectively immobilize enzymes uniformly and robustly within the hydrophobic MCF particles. Compared to the conventional stirring method, the novel pressure-driven method not only greatly increased the enzyme loading, but also significantly reduced the enzyme leaching over multiple reaction cycles. Lipase CALB immobilized on hydrophobic MCF particles by the pressure-driven method showed excellent activity and selectivity in the kinetic resolution of 1-phenylethanol. It also demonstrated superior reusability and thermal stability compared to commercial Novozyme 435 catalyst. CALB/C_{18}-MCF prepared by the pressure-driven method was also successfully applied in a packed bed reactor for continuous reaction with high activity and selectivity.

References

1. (a) Drauz, K.; Waldmann, H. *Enzyme Catalysis in Organic Synthesis: A Comprehensive Handbook*, 2nd ed.; Wiley-VCH: Weinheim, Germany, **2002**; Vol. 1–3. (b) Straathof, A. J. J.; Panke, S.; Schmid, A. *Curr. Opin. Biotechnol.* **2002**, *13*, 548. (c) Liese, A.; Seelbach, K.; Wandrey, C. *Industrial Biotransformations*; Wiley-VCH: Weinheim, Germany, **2000**.
2. (a) Luckarift, H. R.; Spain, J. C.; Naik, R. R.; Stone, M. O. *Nat. Biotechnol.* **2004**, *22*, 211. (b) Lei, C.; Shin, Y.; Liu, J.; Ackerman, E. J. *J. Am. Chem. Soc.* **2002**, *124*, 11242. (c) Bornscheuer, U. T. *Angew. Chem. Int. Ed.* **2003**, *42*, 3336.
3. (a) Mitchell, D. T.; Lee, S. B.; Trofin, L.; Li, N.; Nevanen, T. K.; Soderlund, H.; Martin, C. R. *J. Am. Chem. Soc.* **2002**, *124*, 11864. (b) Kumar, C. V.; McLendon, G. L. *Chem. Mater.* **1997**, *9*, 863. (c) Corma, A.; Fornés, V.; Jordá, J. L.; Rey, F.; Fernandez-Lafuente, R.; Guisan, J. M.; Mateo, C. *Chem. Commun.* **2001**, 419. (d) Wang, P.; Dai, S.; Waezsada, S. D.; Tsao, A. Y.; Davison, B. H.

Biotechnol. Bioeng. **2001**, *74*, 249. (e) Dyal, A.; Loos, K.; Noto, M.; Chang, S. W.; Spagnoli, C.; Shafi, K. V. P. M.; Ulman, A.; Cowman, M.; Gross, R. A. *J. Am. Chem. Soc.* **2003**, *125*, 1684.

4. (a) Takahashi, H.; Li, B.; Sasaki, T.; Miyazaki, C.; Kajino, T.; Inagaki, S. *Chem. Mater.* **2000**, *5*, 465. (b) Yiu, H. H. P.; Botting, C. H.; Botting, N. P.; Wright, P. A. *Phys. Chem. Chem. Phys.* **2001**, *3*, 2983. (c) Blanco, R. M.; Terreros, P.; Fernández-Pérez, M.; Otero, C.; Díaz-González, G. *J. Mol. Catal. B: Enzym.* **2004**, *30*, 83. (d) Yiu, H. H. P.; Wright, P. A.; Botting, N. P. *J. Mol. Catal. B: Enzym.* **2001**, *15*, 81. (e) Dumitriu, E.; Secundo, R.; Patarin, J.; Fechete, I. *J. Mol. Catal. B: Enzym.* **2003**, *22*, 119. (f) He, J.; Li, X.; Evans, D. G.; Duan, X.; Li, C. *J. Mol. Catal. B: Enzym.* **2000**, *11*, 45. (g) Wang. Y.; Caruso, F. *Chem. Mater.* **2005**, *17*, 953.
5. Fernández-Lafuente, G.; Terreni, M.; Mateo, C.; Bastida, A.; Fernández-Lafuente, R.; Dalmases, P.; Huguet, J.; Guisan, J. M. *Enzyme Microb. Technol.* **2001**, *28*, 389.
6. (a) Reetz, M. T.; Simpelkamp, J.; Zonta, A. German Patent DE 4408152 A1, **1995**. (b) Reetz, M. T. *Adv. Mater.* **1997**, *9*, 943. (c) Maury, S.; Buisson, P.; Pierre, A. C. *J. Mol. Catal. B: Enzym.* **2002**, *19*, 269. (d) Reetz, M. T.; Tielmann, P.; Zonta, A. *Adv. Synth. Catal.* **2003**, *345*, 717. (e) Veum, L.; Hanefeld, U.; Pierre A. *Tetrahedron* **2004**, *60*, 10419. (f) Bhatia, R. B.; Brinker, C. J.; Ashley, C. S.; Harris, T. M. *Mater. Res. Soc. Symp. Proc.* **1998**, *519*, 183.
7. (a) Lalonde, J.; Margolin, A. *Enzyme Catalysis in Organic Synthesis*; Wiley-VCH: Weinheim, Germany, **2002**, Vol. 2, p. 163. (b) Cao, L.; van Rantwijk, F.; Sheldon, R. A. *Org. Lett.* **2000**, *2*, 1361. (c) Yazbeck, D. R.; Martinez, C. A.; Hu, S.; Tao, J. *Tetrahedron: Asymmetry* **2004**, *15*, 2757.
8. Arroyo, M.; Sanchez-Montero, J. M.; Sinisterra, J. V. *Enzyme Microb. Technol.* **1999**, *24*, 3.
9. (a) Lettow, J. S.; Han, Y. J.; Schmidt-Winkel, P.; Yang, P.; Zhao, D.; Stucky, G. D.; Ying, J. Y. *Langmuir* **2000**, *16*, 8291. (b) Schmidt-Winkel, P.; Lukens, W. W., Jr.; Yang, P.; Margolese, D. I.; Lettow, J. S.; Ying, J. Y.; Stucky, G. D. *Chem. Mater.* **2000**, *12*, 686. (c) Lettow, J. S.; Lancaster, T. M.; Glinka, C. J.; Ying, J. Y. *Langmuir* **2005**, *21*, 5738.
10. Fan, J.; Yu, C.; Gao, F.; Lei, J.; Tian, B.; Wang, L.; Luo, Q.; Tu, B.; Zhou, W.; Zhao, D. *Angew. Chem. Int. Ed.* **2003**, *42*, 3146.
11. Han, Y.; Lee, S. S.; Ying, J. Y. **2007**, *19*, 2292
12. Lukens, W. W., Jr.; Schmidt-Winkel, P.; Zhao, D.; Feng, J.; Stucky, G. D. *Langmuir* **1999**, *15*, 5403.

366d Understanding Guest-Host Interactions in Sol-Gel Materials

Hugh O'Neill, Volker Urban, and Guangming Luo. Chemical Sciences Division, Oak Ridge National Laboratory, PO Box 2008, Oak Ridge, TN 37831

Living cells are often described as crowded environments because macromolecules occupy a large volume of the fluid. In these systems non-specific interactions are individually weak but cumulatively strong. One particular type that arises from steric repulsion is excluded volume interactions that can increase the free energy of a solution and the chemical potential of the solutes. The same interactions that play a role in the cellular environment are also important for investigation of the interactions at the biotic-abiotic interface. The properties of proteins were investigated in an artificially crowded environment formed by their entrapment in sol-gels. The influence of surface chemistry on the properties of the entrapped biomolecules and their distribution in the sol-gel matrix was investigated. Data will be presented on how sol-gel matrices functionalized with aliphatic, ionic, and uncharged hydrophilic groups affected the folding properties of proteins in sol-gels. These studies were complimented with small angle scattering and fluorescence techniques to investigate the distribution of the entrapped proteins. These studies are important for enzyme based-diagnostic and biosensing applications.

366e Activation of Enzymes in Organic Solvents by Immobilization on Silica Nanoparticles

Juan C. Cruz, **Peter Pfromm**, and Mary E. Rezac. Chemical Engineering, Kansas State University, Manhattan, KS 66506

The significant activation of subtilisin Carlsberg for catalysis in hexane by immobilization on fumed silica nanoparticles has been shown by us previously. Since the results for this activation were similar to the best published activation method at the time (lyophilization in presence of inorganic salts) we were interested if fumed silica activation can be generalized to other enzymatically catalyzed reactions in solvents. Here we apply fumed silica activation to a model esterification catalyzed by crude C. Antarctica Lipase B in hexane and we compare to a commercial preparation of the same enzyme (Novozyme 435). The activity of our fumed silica activated enzyme preparation for the model esterification appears on the same order as that of Novozyme 435 (normalized by active units of enzyme). The activity also depends on the amount of silica available for protein immobilization. It appears that fumed silica activation for enzymatic catalysis in solvents can be applied to enzymes other than S. Carlsberg. This appears to support our previous hypothesis that the area available for enzyme immobilization is crucial for enzyme activation.

366f Nanoparticle-Supported Multi-Enzyme Biocatalysis

Wen-Fang Liu[1], Mu-Qing Zheng[2], Song-Ping Zhang[2], and **Ping Wang**[3]. (1) Institute of Process Engineering, Chinese Academy of Sciences, Beijing, 100080, China; Beijing Institute of Technology, Beijing, 100081, China, (2) Institute of Process Engineering, Chinese Academy of Sciences, Beijing, 100080, China, (3) Department of Bioproducts and Biosystems Engineering, University of Minnesota, St Paul, MN 55108

Glutamate dehydrogenase (GLDH), lactate dehydrogenase (LDH) and coenzyme NAD+ were attached onto silica nanoparticles examined for biosynthesis of α-ketoglutarate from L-glutamate. The immobilized multi-enzyme catalyst retained about 50% of the activity of the free-enzyme system for the coupled reactions. Since nanoparticles always tend to agglomerate in the absence of dispersing agents such as surfactants, it was found that liquid-solid ratio of the reaction system had a great effect on the activity of the catalyst system. Overall, the particle-attached NAD^+ showed effective interaction with the enzymes immobilized in the same way and achieved a TTN of around 20000 within one hour of reaction, corresponding to a glutamate conversion of 22.7%. Magnetic nanosilica particles were also examined for the immobilization of the catalyst system for easy catalyst recovery and reuse.

366g Silica Formation and Acid Phosphatase Immobilization Using Small Bi-Functional Biomimetic Catalysts at near Neutral pH Conditions

Herminio C. De Sousa, Chemical Engineering Department, University of Coimbra, Polo II - Pinhal de Marrocos, Rua Silvio Lima, Coimbra, 3030-790, Portugal, David M. Hess, Department of Chemical Engineering, University of Massachusetts, 51 Goessmann, Amherst, MA 01002, and James J. Watkins, Department of Polymer Science and Engineering, University of Massachusetts, 120 Governors Dr., Amherst, MA 01003.

Several marine organisms, including sponges and diatoms, can produce solid silica structures with well defined nanoscale morphologies, at ambient and neutral pH by peptide mediated condensation of naturally occurring silicic acid.1,2 To date, a number of groups have studied biosilica formation in the laboratory and have successfully developed biological mimics for the condensation of disordered silica structures at ambient conditions.3-6 More recently, polyamines and small bifunctional molecules, including cysteamine and ethanolamine, have been shown to be effective biomimetic catalysts for silica formation.7-10 An application of silica condensation at ambient conditions is the silica immobilization of labile biomolecules (including enzymes, proteins and drugs) which in contrast to most conventional sol-gel routes, would avoid their possible degradation and/or loss of their bioactivity under harsh immobilization conditions.11-14

In this work, two small by-functional catalysts (cysteamine and methionine), were employed as biomimetic catalysts for the simultaneous mesoporous silica formation and acid phosphatase immobilization from tetraethyl orthosilicate (TEOS). Near neutral pH and near physiological temperature conditions were used. The enzyme-loaded silica samples were characterized and quantified using a variety of techniques. Immobilized acid phosphatase activity was determined and compared to free enzyme activity. Results indicated that acid phosphatase entrapped in precipitated silica retained its enzymatic activity. Effects of temperature and catalysts' concentration on silica formation yields and on enzyme activity and immobilization yields were investigated and will be discussed.

References

(1) Lewin, J.C., "Silicon metabolism in diatoms. I. Evidence for the role of reduced sulfur compounds in silicon utilization", J. Gen. Physiol., 37, 589-599, 1954.

(2) Mann, S., "Biomineralization and biomimetic materials chemistry", J. Mater. Chem., 5(7), 935-946, 1995.

(3) Naik, R.R., Whitlock, P.W., Rodriguez, F., Brott, L.L., Glawe, D.D., Clarson, S.J., Stone, M.O., "Controlled formation of biosilica structures in vitro", Chem. Comm., 2, 238-239, 2003.

(4) Poulsen, N., Sumper, M., Kroger, N., "Biosilica formation in diatoms: characterization of native

silafin-2 and its role in silica morphogenesis", Proc. Natl. Acad. Sci. USA, 100(21), 12075-12080, 2003.

(5) Shimizu, K., Cha, J.N., Stucky, G.D., Morse, D.E., "Silicatein alpha: Cathepsin L-like protein in sponge biosilica", Proc. Natl. Acad. Sci. USA, 95, 6234-6238, 1998.

(6) Cha, J.N., Shimizu, K., Zhou, Y., Christiansen, S.C., Chmelka, G., Stucky, G.D., Morse, D.E., "Silicatein filaments and subunits from a marine sponge direct the polymerization of silica and silicones in vitro", Proc. Natl. Acad. Sci. USA, 96, 361-365, 1999.

(7) Cha, J.N., Stucky, G.D., Morse, D.E., Deming, T.J., "Biomimetic synthesis of ordered silica structures mediated by block copolypetides", Nature, 403, 289-292, 2000.

(8) Roth, K.M., Zhou, Y., Yang, W., Morse, D.E., "Bifunctional small molecules are biomimetic catalysts for silica synthesis at neutral pH", J. Am. Chem. Soc., 127(1), 325-330, 2005.

(9) Belton, D.J., Patwardhan, S.V., Perry, C.C., "Spermine, spermidine and their analogues generate tailored silicas", J. Mater. Chem., 15,4629-4638, 2005.

(10) Sumper, M., Kroger, N., "Silica formation in diatoms: the function of long-chain polyamines and sillafins", J. Mater. Chem., 14, 2059-2065, 2004.

(11) Gill, I., Ballesteros, A., "Encapsulation of biologicals within silicate, siloxane, and hybrid sol-gel polymers: an efficient and generic approach", J. Am. Chem. Soc., 120, 8587-8598, 1998.

(12) Gill, I., "Bio-doped nanocomposite polymers: sol-gel bioencapsulates", Chem. Mater., 13, 3404-3421, 2001.

(13) Luckarift, H.R., Spain, J.C., Naik, R.R., Stone, M.O., "Enzyme immobilization in a biomimetic silica support", Nature Biotechn., 22(2), 211-213, 2004.

(14) Avnir, D., Coradin, T., Lev, O., Livage, J., "Recent bio-applications of sol-gel materials", J. Mater. Chem., 16, 1013-1030, 2006.

366h Enzyme Encapsulation in Oxide Matrices

Heather R. Luckarift, D. Matthew Eby, Lloyd.J Nadeau, and Glenn R. Johnson. AFRL/MLQL, Air Force Research Laboratory, 139 Barnes Drive, Suite # 2, Tyndall Air Force Base, FL 32403

Biomineralization reactions provide an effective mechanism to generate inorganic matrices that are suitable for enzyme immobilization. A range of cationic proteins and polypeptides, for example, catalyze the biomineralization of silica in vitro; the biomolecules acting as a scaffold during the formation of the silica matrix. In general the silica precipitation reaction forms a network of fused spherical silica nanoparticles, although size and morphology is dependent upon the choice of catalyst and the reaction environment. When additional enzymes are added during the formation of silica, the enzymes become encapsulated within the silica particles as they form. The silica forms at ambient environmental conditions, providing mild encapsulation conditions and high biocompatibility to the encapsulated biomolecules. The versatility of the silica-immobilization method has been demonstrated for a variety of enzymes with consistently high immobilization efficiencies and retention of enzyme activity. The versatility of silica-encapsulated enzymes will be presented for a range of model applications including biocatalysis, development of biosensors and applications to biofuel cell design. In addition, lysozyme has recently been shown to act as a scaffold for silica formation. Lysozyme retains its anti-microbial properties when entrained within the resulting silica particles. Advances in the characterization and elucidation of these mechanisms in respect to the development of anti-microbial coatings will also be presented.

397a Past, Present, and Future Opportunities for Bionanotechnology in Cbet

Robert Wellek, The National Science Foundation, Division of Chemical, Bioengineering, Environmental and Transport Systems, 4201 Wilson Boulevard, Arlington, VA 22230

This presentation will give an overview of past, present, and future opportunities for bionanotechnology in the Chemical, Bioengineering, Environmental, and Transport Systems (CBET) Division of the National Science Foundation (NSF).

433a Nanopore Engineering of Chitosan Polymer for Enzyme Immobilization and Stabilization

Shelley D. Minteer[1], Tamara Klotzbach[1], Michael Cooney[2], and Bor Yann Liaw[2]. (1) Chemistry, Saint Louis University, 3501 Laclede Ave., St. Louis, MO 63103, (2) Hawaii Natural Energy Institute, SOEST, University of Hawaii, 1680 East West Road, POST 109, Honolulu, HI 96822

This research details the development of a hydrophobically modified chitosan membrane for enzyme immobilization and stabilization. Chitosan is a linear polysaccharide that is formed from chitin, which is abundant in the outer shells of insects and crustaceans. Chitosan is an inexpensive biopolymer, because it is the second most abundant biomass. Well known for it biocompatibility, biodegradability, chemical inertness, high mechanical strength, and non-toxic properties, chitosan is widely used in medical and biosensor applications. Chitosan was modified by reacting a saturated alkyl aldehyde with the free amino group on the chitosan backbone. This hydrophobic modification induces a micellar structure in the chitosan resulting in the formation of a mesopore structure whose size, pH, and hydrophobicity can be engineered for enzyme immobilization and stabilization. Studies of the physical and chemical properties of the membranes were completed, along with studies of the immobilization of dehydrogenase enzymes, including: NAD-dependent glucose dehydrogenase, alcohol dehydrogenase, aldehyde dehydrogenase, lactic dehydrogenase, and formate dehydrogenase. Lifetime studies have shown enzyme activity is maintained for greater than 6 months within the hydrophobically modified chitosan membrane, promising widespread application in biosensor and biofuel cell development.

433b Application of Mesopore Engineered Chitosan Polymer for Fabrication of Multi-Dimensional and Multi-Directional Enzyme Catalyzed Electrodes

Michael Cooney[1], Bor Yann Liaw[1], Carolin Lau[1], and Shelley D. Minteer[2]. (1) Hawaii Natural Energy Institute, SOEST, University of Hawaii, 1680 East West Road, POST 109, Honolulu, HI 96822, (2) Chemistry, Saint Louis University, 3501 Laclede Ave., St. Louis, MO 63103

Porous organic polymer matrices possess promise as a biofuel cell enabling strategy, primarily because of their ability to provide multi-dimensional and multi-directional pore structures that can improve performance in terms of power density and enzyme lifetime - two major impediments to any viable biofuel cell application. In the previous talk the capacity to hydrophobically pretreat chitosan polymer to possess hydrophobic mesopores that both immobilize and stabilize enzymes has been discussed. In this talk, the fabrication of multidimensional and multidirectional macroporous chitosan/enzyme electrodes from the hydrophobically pretreated chitosan is presented. The purpose of the macropores is to provide channels for mass transport of liquid phase fuels, thereby enabling the fabrication of 3D electrodes rather than thin films. Data on the electrochemical performance of these electrodes will be presented along with commentary on their potential to improve power density.

Aligned Carbon Nanotubes for Bio-sensing and Bio-fuel Cell Applications

Kuanping Gong, Lingti Qu, Pinggang He and Liming Dai

Department of Chemical and Materials Engineering, School of Engineering and UDRI, University of Dayton, 300 College Park, Dayton, OH 45469-0240, USA

The recent development of nanotechnology has opened up novel fundamental and applied frontiers in materials science and engineering.[1] Due to their high surface-to-volume ratio and unique electronic properties, carbon nanotubes and their derivatives have been widely studied for chemical-/bio-sensors and bio-fuel cell applications, among many others.

Since the early demonstrations of the ability of carbon nanotubes for gas sensing by charge transfer with the gas molecules,[2,3] the transduction mechanism has been significantly extended to include the electrochemical detection for chemical-/bio-sensing.[4] A large number of biosensors employ enzymatic catalysts that react with the analyte producing an electrochemically active species, such as hydrogen peroxide.[5] An electrochemical method is then used to detect the electrochemically active species quantifying the analyte presented. For these biosensors, large electrode surfaces of the sensor array and good conductivity of the electrode and electrolyte will enhance significantly their performance. Thus, the high surface area and good electronic property provided by carbon nanotubes is an attractive feature in the advancement of a chemical or biosensor. In most of the earlier studies, however, nonaligned carbon nanotubes were used.[4,5] It will be a significant advancement for a carbon nanotube based sensor, if we can prepare perpendiculary-aligned carbon nanotubes as the sensing material.[6] This is because aligned carbon nanotubes will provide a well-defined large surface area for a sensor with an enhanced sensitivity. These sensors can be built on each of the constituent aligned carbon nanotubes while they can be addressed collectively through the common electrode. The alignment structure will also facilitate surface modification of the carbon nanotubes with various transduction materials for broadening the scope of analytes to be detected by the nanotube sensor. Micropatterning of the aligned carbon nanotubes will further allow for the development of nanotube sensor chips.

We have previously prepared aligned carbon nanotubes with a well-graphitized structure, having *ca.*50 concentric carbon shells and an outer diameter of *ca.*40 nm, by pyrolyzing iron(II) phthalocyanine (FePc) under Ar/H_2 atmosphere at 800-1100°C.[7] Subsequently, we have also developed microfabrication methods for patterning the aligned carbon nanotubes with a sub-micrometer resolution and for patterned/non-patterned transferring such nanotube arrays to various other substrates of particular interest (*e.g.* polymer films for organic optoelectronic devices or metal substrates for electrochemistry).[7-11] Using the aligned carbon nanotubes as the sensing material, we have recently developed new sensors of a high sensitivity, good selectivity, and excellent environmental stability for the detection of a broad class of chemical vapors with low power consumption by partially coating of perpendicularly-aligned carbon nanotube arrays with an appropriate polymer thin film along their tube length.[12] As schematically shown in Figure 1, the absorption and desorption of chemical vapors by the polymer matrix cause changes in the inter-tube distance, and hence the surface resistance across the nanotube film. Simple measurements of the resistance change, therefore, constituent the nanotube-polymer chemical vapor sensors.[12]

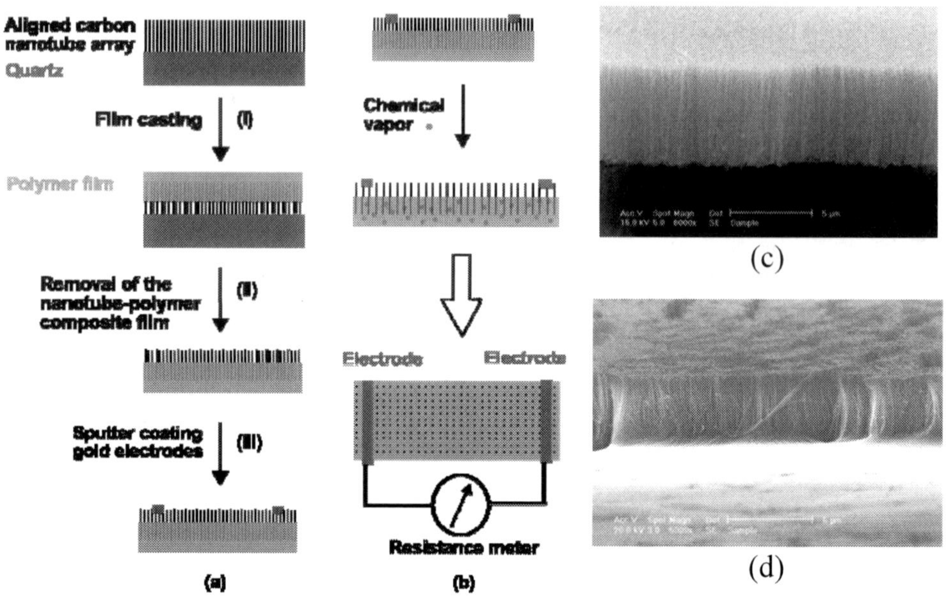

Figure 1. Schematic illustration of the procedures for (a) fabricating and (b) characterizing the aligned carbon nanotube-polymer composite chemical vapor sensor. SEM images of the aligned carbon nanotube array (c) before and (d) after being partially coated with a polymer (PVAc) film on top and turned upside down (Scale bars: 5 μm).[12]

We have also developed a simple, but effective, method for preparing aligned carbon nanotube–DNA sensors by chemically coupling DNA probes on both the tip and wall of plasma-activated aligned carbon nanotubes in a patterned and non-patterned fashion.[13] In addition to the chemical grafting of polymer chains onto the carbon nanotube surface, we have further used the aligned carbon nanotubes as nanoelectrodes for making novel conducting coaxial nanowires by electrochemically depositing a concentric layer of an appropriate conducting polymer uniformly onto each of the constituent aligned nanotubes to form the aligned conducting polymer coated carbon nanotube coaxial nanowires (CP-NT).[14] To demonstrate the potential sensing applications for the CP-NT coaxial nanowires, we have also immobilized glucose oxidase onto the aligned carbon nanotube substrate by electropolymerization of pyrrole in the presence of glucose oxidase. The glucose oxidase-containing polypyrrole-carbon nanotube coaxial nanowires were used to monitor the concentration change of hydrogen peroxide (H_2O_2) generated from the glucose oxidation reaction by measuring the increase in the electro-oxidation current at the oxidative potential of H_2O_2 (*i.e.* the amperometric method).[15] With these CP-CNT nanowire sensors, we observed a linear glucose-concentration-dependence for the electrooxidation current, which typically is about ten times higher than that of more conventional flat electrodes coated with glucose oxidase-containing polypyrrole films under the same conditions. Owing to its high selectivity and activity toward H_2O_2 reduction, prussian blue (PB, $KFe[Fe(CN)_6]$ or $Fe_4[Fe(CN)_6]_3$) has also been immobilized onto our aligned carbon nanotube sensors as electrocatalysts to further facilitate the electron communication between the analyte and substrate electrode.[16]

Recently, we have developed a substrate-enhanced electroless deposition (SEED) method to assemble metal nanoparticles onto either aligned or nonaligned carbon nanotubes.[17,18] These carbon nanotubes decorated with various metal nanostructures were demonstrated to possess interesting optical and electrocatalytic properties for developing novel sensors. More recently, we have also prepared multicomponent and multifunctional hybrid structures based on microsized carbon fibers sheathed with aligned carbon nanotubes.[19] Based on our previously-reported dry contact transfer technique,[11,20] we have successfully constructed multidimensional and multifunctional nanomaterials with the microsized carbon fibers sheathed by patterned/nonpatterned aligned carbon nanotubes along their length.[11] We have further used the SEED method to modify the carbon-fiber-supported aligned carbon nanotubes with metal nanoparticles and other substances. The resultant microsized carbon fibers sheathed with aligned carbon nanotubes and their derivatives provide a useful means to connect nanoscale entities to the outside world and were shown to possess interesting electrochemical properties attractive for a wide range of potential applications, including in highly sensitive chemical and bio-sensors and methanol direct fuel cells.

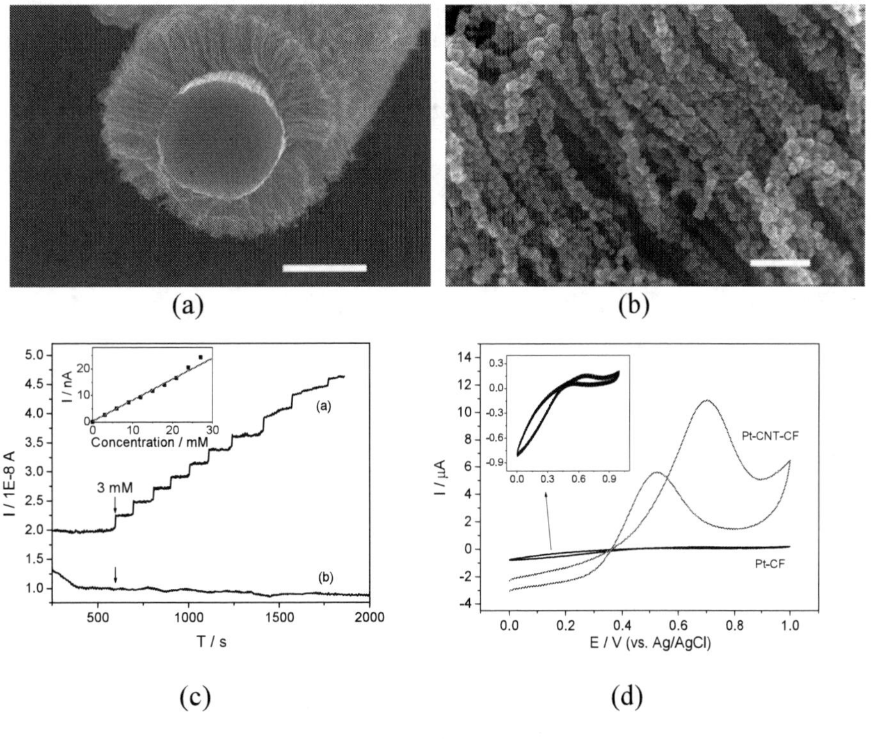

Figure 2. SEM images of (a) a cross-sectional view of aligned carbon nanotubes surrounding the carbon fiber along its length (scale bar: 5 μm) and (b) Pt nanoparticles deposited on the carbon-fiber-supported aligned carbon nanotubes (scale bar: 500 nm). (c) Amperometric responses to the successive additions of 3 mM glucose for a glucose-oxidase-attached aligned carbon nanotubes supported by the carbon fiber electrode. (d) Cyclic voltammograms the Pt-nanoparticle-attached aligned carbon nanotubes supported by the carbon fiber in an aqueous solution of 2 M CH_3OH/0.1 M H_2SO_4 at a scan rate of 50 mV/s. The inset shows an enlarged view of the CV for a reference electrode with Pt nanoparticles attached onto a pure carbon fiber.[19]

Just as aligned carbon nanotubes have been demonstrated to show excellent sensing performance, the high surface area and good electronic property provided by carbon nanotubes is also an attractive feature in the advancement of bio-fuel cells. In this context, we have developed an alkaline fuel cell by using glucose as fuel and aligned carbon nanotube arrays as electrodes. The assembled fuel cell showed good performance with an open circuit voltage of 0.55 V and output power density of 48.0 μW at 0.35 V, indicating that aligned carbon nanotubes could be a potential alternative to the commercial Pt-loaded carbon cathodes for fuel cell applications. [16.]

In this talk, we will present an overview on the potential use of aligned carbon nanotubes in chemical-/bio-sensors and bio-fuel cells by spotlighting some of the examples described above. The potential use of these aligned carbon nanotubes for other related applications will also be discussed.

References

1. Dai, L. (Ed.), *Carbon Nanotechnology: Recent Developments in Chemistry, Physics, Materials Science and Device Applications*, Elsevier: Amsterdam, 2006.
2. Kong, J.; Franklin, N.R.; Zhou, C.; Chapline, M.G.; Peng, S.; Cho, K.; Dai, H. *Science* **2000**, *287*, 622.
3. Collins, P.G.; Bradley, K.; Ishigami, M.; Zettl, A. *Science* **2000**, *286*, 1801.
4. Ohashi, T.; Dai, L. in 'Nanotechnology: Recent Developments in Chemistry, Physics, Materials Science and Device Applications' (Ed. Dai, L.); Elsevier: Amsterdam, 2006.
5. He, P.; Dai, L. in '*Biomedical and Biological Nanotechnology*' (Eds. Lee, J.; Lee, A.); Volume 1 of *The Handbook of Biomems and Bio-nanotechnology* (Ed. Ferrari, M.); Kluwer Academic Publishers: Amsterdam, 2005.
6. Dai, L.; Patil, A.; Gong, X.; Guo, Z.; Liu, L.; Liu, Y.; Zhu, D.B. *ChemPhysChem* **2003**, *4*, 1150.
7. Huang, S.; Dai, L.; Mau, A.W.H. *J. Phys. Chem. B* **1999**, *103*, 4223.
8. Y. Yang, S. Huang, H. He, A.W.H. Mau, L. Dai, *J. Am. Chem. Soc.* **1999**, *121*, 10832, and references cited therein.
9. Patil, A.; Ohashi, T.; Buldum, A.; Dai, L. *Appl. Phys. Lett.* **2006**, *89*, 103103.
10. Huang, S.; Mau, A.W.H.; Turney, T.W.; White, P.A.; Dai, L. *J. Phys. Chem. B* **2000**, *104*, 2193.
11. Yang, J.; Qu, L.; Zhao, Y.; Zhang, Q.; Dai, L.; Baur, J.W.; Maruyama, B.; Vaia, R.A.; Shin, E.; Murray, P.T.; Luo, H.; Guo, Z.-X. *J. Nanosci. Nanotechnol.* **2007**, *7*, 1573.
12. Wei, C.; Dai, L.; Roy, A.; Benson Tolle, T. *J. Am. Chem. Soc.* **2006**, *128*, 1412.
13. He, P.; Dai, L. *Chem. Commun.* **2004**, 348.
14. Gao, M.; Huang, S.; Dai, L.; Wallace, G.; Gao, R.; Wang, Z. *Angew. Chem. Int. Ed.* **2000**, *39*, 3664.
15. Gao, M.; Dai, L.; Wallace, G. *Electroanalysis* **2003**, *15*, 1089.
16. Gong, K.P; Dai, L. (to be published).
17. Qu, L.; Dai, L. *J. Am. Chem. Soc.* **2005**, *127*, 10806.
18. Qu, L.; Dai, L.; Ōsawa, E. *J. Am. Chem. Soc.* **2006**, *128*, 5523, references cited therein.
19. Qu, L.; Zhao, Y.; Dai, L. *Small* **2006**, *2*, 1052.
20. Yang, J.; Dai, L.; Vaia, R.A. *J. Phys. Chem. B* **2003**, *107*, 12387.

Acknowledgement. We thank our colleagues, particularly Wei Chen, Mei Gao, Gordon Wallace, Ajit Roy, and Tia Benson-Tolle for their contributions to the work summarized in the article. P.H. is on leave of absence from Department of Chemistry, East China Normal University, Shanghai 200062, China. L.D. is grateful for financial support from the NSF (CMMI-0708055) and AFOSR (FA9550-06-1-0384).

433d Simulation of Spacially-Resolved, Multistep Bioelectrocatalytic Processes

Scott Calabrese Barton, Chemical Engineering and Materials Science, Michigan State University, East Lansing, MI 48824-1226

Multistep enzymatic cascades may be applied to the oxidation of fuels to yield high fuel-efficiency electrochemical energy systems. The ability to structure and pattern anodes at the micron- to nano-scale enables precise placement and orientation of electrodes to maximize utilization of stable intermediates and minimize byproducts. We will present the results of a theoretical analysis of the multistep oxidation of methanol and glucose fuels via enzyme catalysis in which each enzymatic step is resolved to an individual, nanoscale electrode. These results are compared to those obtained for single electrodes with randomly mixed biocatalysts, and to experimental results obtained in a microfluidic biofuel cell.

433e An Enzyme-Nanofiber Composite for Stable and Continuous Long-Term Enzymatic Reaction

Byoung-In Sang[1], Jin Hyung Lee[2], Ee Taek Hwang[3], Byoung Chan Kim[2], Sun-Mi Lee[1], Jungbae Kim[4], and Man Bock Gu[5]. (1) Center for Environmental Technology Research, Korea Institute of Science and Technology, Seoul, South Korea, (2) Advanced Environmental Monitoring Research Center (ADEMRC), Gwangju Institute of Science and Technology (GIST), 1, Oryoung-dong, Puk-Gu, Gwangju, South Korea, (3) College of Life Sciences and Biotechnology, Korea University, Seoul, South Korea, (4) Pacific Northwest National Laboratory, 902 Battelle Blvd., PO Box 999, Mailstop K8-93, Richland, WA 99352, (5) Department of Biotechnology, Korea University, Seoul, South Korea

For long-term stable operation of enzymatic degradation process, this study shows the preparation and application of enzyme–nanofiber composites, which were prepared by coating an enzyme aggregate, the esterase from Rhizopus oryzae, on the surface of the nanofibers. After immobilization on the nanofiber, the apparent Km for the immobilized esterase was 1.48-fold higher than that of the free esterase, with values of 0.98 and 1.35 mM for the free and immobilized enzymes, respectively. The activity of enzymes in enzyme–nanofiber was very stable, even when the fibers were shaken in glass vials, preserving 80% of the initial activity for 100 days. In addition, the enzyme–nanofiber composite was used repeatedly in 30 cycles of substrate hydrolysis and still remained active. Consequently, the esterase–nanofiber composite was employed within a continuous reactor system to evaluate its use in a long-term and stable continuous substrate hydrolysis reaction. It was found that the production of p-nitrophenol was stable for at least 400 h. This study demonstrates that the enzyme–nanofiber composite can be used in both repeated-batch mode and a continuous mode for a long-term stable operation.

433f Haem Proteins in Nonaqueous Solvents: Electrochemistry and Biocatalysis

Sarah Hudson, University of Limerick, MS1010, University of Limerick, Limerick, Ireland and **Edmond Magner**, Materials and Surface Science Institute, University of Limerick, MS1010, University of Limerick, Limerick, Ireland.

The use of redox enzymes as biocatalysts has been limited, mainly due to their instability which arises from the need for an exogeneous oxidant, typically hydrogen or alkyl peroxides. We are interested in examining the properties of haem proteins with a view to using these proteins as biosensors and biocatalysts. By examining the surface properties of the immobilisation matrix and of the protein of interest, we have established a systematic means of evaluating the potential of a porous support to immobilise enzymes with retention of catalytic activity. Chloroperoxidase was immobilised on to a mesoporous silicate and re-used 20 times without loss of activity. The electrochemical and biocatalytical properties of hemin and a range of haem proteins, (myoglobin, haemoglobin, cytochrome c, horseradish peroxidase and chloroperoxidase) was then examined in nonaqueous solvents. The thermodynamics of reduction of these proteins indicates that each protein needs to be examined independently, with no discernible trends evident for a particular enyzme in a range of solvents. Wih enzymes such as horseradish peroxidase where the haem is noncovalently bound to the protein, it is apparent that the haem is partially removed from the haem binding pocket.

433g Robust Biocatalytic Single Enzyme Nanogels by Versatile Strategies

Ming Yan, Su Li, Zhixia Liu, and **Zheng Liu**. Department of Chemical Engineering, Tsinghua University, Beijing, China

Biocatalysis has been well regarded as an important and cost-effective technology for the manufacture of fine chemicals, synthesis of pharmaceuticals, biosensors, bioremediation, protein digestion in proteomic analysis, and biofuel cells. Stability and activity are always the most concerned properties of enzymes and limited factors to the vary applications. Here we present a facile in-situ aqueous polymerization procedure to fabricate single enzyme nanogels as robust nanometer scale biocatalysis. By using versatile modification strategies, such as acryloylation of lysine group, introducing vinyl group to glutamic acid and asparagine, control of modification pH, active site protection by inhibitors, different single enzyme nanogels are successfully prepared, which show greatly improved stability to temperature, organic solvent and low pH solution without the compromise of catalytic capability. These nanogels are quite suitable to biocatalytic applications in harsh condition, which are constrainted by small size or high mass-transfer rating reaction or exposed for extreme pH and temperature.

433h Lipase Nanogel for Biodiesel Production

Jun Ge, Jun Wang, Xin Chen, Wei Du, and **Zheng Liu**. Department of chemical engineering, Tsinghua University, Beijing, 100084, China

We report here a two-step procedure preparing lipase nanogel. The first step is to introduce vinyl groups onto enzyme surface and the second step is in-situ polymerization using acrylamide as monomers. The nanogel network, while enhancing the stability of encapsulated enzyme via multiple covalent attachments, generates minor additional resistance for mass transport. The greatly enhanced durability against harsh conditions, e.g. high temperature and the presence of organic solvent, indicates a great potential for industrial process of many kind.

Lipase (EC 3.1.1.3) is extensively studied for potential application in synthesis of surfactants and pharmaceuticals, production of biodiesel, resolution of racemic mixtures, and detergent industry. The production of biodiesel from plant oil can be efficiently catalyzed by lipase in a water-containing system without an organic solvent. However, this process is hindered by methanol induced deactivation of lipase.

In this work, lipase nanogel was prepared using above mentioned procedure. Transmission electron microscope (TEM), size exclusion chromatography (SEC), SDS-PAGE, FT-IR were used to characterize the enzyme nanogel. Using p-Nitrophenylburate as substrate, Km and Vmax of lipase nanogel was 0.29 mM, 51 Umg-1 while their respective counterpart of free lipase was 0.28 mM, 87 Umg-1. The half-life time of native lipase was 50 min at 50°C while almost no visible loss of lipase activity was observed in the 500 min for the nanogel. After incubation in 40% methanol aqueous solution, lipase nanogel preserved about 80% of its hydrolytic activity while native lipase reserved only 10%. All these are of great interest for further exploration of the potential application of lipase nanogel for the production of biodiesel.

489a Biomolecular Motors for Directed Assembly and Hybrid Devices

Henry Hess, Ashutosh Agarwal, Matthew Downs, Isaac Finger, Thorsten Fischer, Yolie Jeune, Parag Katira, Isaac Luria, Elizabeth Mobley, Ajoy Saha, and Robert Tucker. Materials Science and Engineering, University of Florida, 160 Rhines Hall, Gainesville, FL 32611

Biomolecular motors, such as the motor protein kinesin, convert the chemical energy stored in adenosine triphosphate with high efficiency into mechanical work. Their nanoscale dimensions and independence from external connections enables them to act as independent agents in a liquid environment, capable of performing a variety of tasks in nanotechnology, such as directed transport or active assembly and disassembly. The integration of such nanoengines into nanodevices and multifunctional materials raises a host of intriguing engineering questions, some related to the biological origin of the motors and others of general relevance to the field of molecular motors. One example is the balance between external control and self-organization in these multi-agent systems. Other questions are related to the achievable gains in required energy, the limits of power density, or the prediction of device and material characteristics. Finally, fundamental research has to be complemented by directed work towards applications. Our increasing experience with the integration of biomolecular motors into synthetic devices and the expanding knowledge about the biological functions of motor proteins sharpen the focus on the uniqueness and feasibility of application ideas related to, for example, biosensors and advanced materials.

489b A Novel Self-Assembled Protein Nanostructure as Multifunctional Catalyst for Xylan Hydrolysis

Shara McClendon[1], Hyun-Dong Shin[1], Zichao Mao[2], and **Rachel Ruizhen Chen**[3]. (1) Chemical and Biomolecular Engineering, Georgia Institute of Technology, 778 Atlantic Dr, Atlanta, GA 30332, (2) School of Chemical and Biomolecular Engineering, Georgia Institute of Technology, 311 Ferst Dr. NW, Atlanta, GA 30332-0100, (3) Chemical&Biomolecular Engineering, Georgia Institute of Technology, 311 Ferst Drive, NW, Atlanta, GA 30332-0100

Cellulosome, a multi-protein nanostructure (18 nm), is an active nanosystem naturally made by many anaerobic microbes to assimilate recalcitrant cellulose. This self-assembled system brings multiple enzymes to proximity of the substrate, and provides a structure that ensures a high local concentration and correct ratio/orders of the components, thereby maximizing synergy. Consequently, it has much higher catalysis efficiency than soluble enzymes present in a non-organized fashion. The molecular architecture of a cellulosome consists of a structural scaffold protein (or scaffoldin) containing multiple repeating units called cohesins, into which multiple enzymes, each tagged with a complementary end (called dockerin), can dock and self-assemble into a multi-protein complex. The specific protein-protein, or complementary cohesin-dockerin interaction, provides the mechanism for position-specific self-assembly.

Using cellulosome architecture as design template, we have designed a two-unit nano protein structure, named xylosome, for xylan hydrolysis. The xylosome consists of a two-cohesin scaffold protein and two xylan-degrading enzymes (a xylanase and a xylosidase) each tagged with a dockerin. Xylanase from Clostridium thermocellum with a dockerin complementary to the first cohesin unit was cloned from and overexpressed in E. coli. Other components of the nanostructure are being constructed with gene fusion strategy. These components, after isolation and purification, will be allowed to self-assemble. A range of environmental conditions, such as pH, ionic strength, calcium concentrations, temperatures, the presence of other proteins, and target protein concentration, will be investigated in order to understand how they influence self-assembly, stability, and functionality of the nanostructure. The performance of xylosome will be evaluated with xylan substrates.

The presentation will detail the molecular design, construction, and characterization of this novel protein nanostructure and highlight its application as a high-performance multifunctional catalyst in hydrolysis of xylan.

489c Development of a Nanoparticle Based Nanosome to Enhance Enzyme Proximity Synergy between β-Glucosidase and Cellobiohydrolase

Devesh Srivastava[1], Shishir P. S. Chundawat[2], Bruce Dale[3], and Ilsoon Lee[3]. (1) Chemical Engineering and Materials Science, Michigan State University, 2527 Engineering Building, East Lansing, MI 48824, (2) Department of Chemical Engineering and Materials Science, Michigan State University, Room 3263 Engineering Building, East Lansing, MI 48824-1226, (3) Michigan State University, Department of Chemical Engineering and Materials Science, Room 3247 Engineering Building, East Lansing, MI 48824-1226

Certain cellulolytic microbes (i.e., Clostridium, Acetivibrio, Bacteroides, and Ruminococcus) produce an intricate self-assembly of multi-enzyme complexes, called cellulosomes, which hydrolyze cellulosic and hemicellulosic biomass. Cellulosomes essentially consist of a non-enzymatic scaffolding protein associated with several enzyme subunits that act in concert to degrade biomass. There has been recent interest in constructing designer mini-cellulosomes, or nanosomes, to enhance synergistic interaction between enzyme components.

Cellobiohydrolase (CBH) and β-glucosidase (BG) are two important classes of glucosyl hydrolases that have significant synergistic interaction between each other. However, the interaction between the two enzymes is diffusion limited. One way of further improving the proximity synergistic action of CBH and BG is to construct nanosomes with the two enzymes co-immobilized. Nanoparticles such as carbon nanotubes (CNT) and polymeric nanoparticles, provide a suitable scaffold whose surface properties can be modified to immobilize enzymes. We have developed a nanosome with co-immobilized β-glucosidase and cellobiohydrolase. The enzymatic activity of the nanosome is measured using soluble (i. e., oligosaccharides) and insoluble substrates (i.e., crystalline cellulose). The nanosome is further studied using transmission electron microscopy (TEM) and X-ray photoelectron spectroscopy (XPS). Enzyme deposition on nanoparticles was further characterized by fluorescence confocal microscopy and change in the zeta (ζ) potential of nanoparticles due to enzyme adsorption on the surface. Enzyme loading was determined using UV-absorption spectroscopy.

489d A Novel Approach to Ultrasensitive Detection of Disease Marker Using Protein Nanoparticles

Hyewon Lee, Sung-Hyun Lee, Jin-Seung Park, and **Jeewon Lee**. Department of Chemical and Biological Engineering, Korea University, Anam-Dong 5-1, Seoul, South Korea

We report on the ultrasensitive protein nanoprobe system that specifically captures disease marker (autoantibodies of Type I diabetes in this case) with attomolar sensitivity. The system relies on supramolecular protein nanoparticles that bind the early marker of Type I diabetes. The ultrasensitive detection of early marker of Type I diabetes during the early phase of pancreatic â-cell destruction is important because individuals at high risk of developing Type I diabetes can be identified several years before the clinical onset of the ailment. The bacterial expression of chimera genes encoding N-[human ferritin heavy chain (hFTN-H)]::[specific antigenic epitope]-C produces supramolecular nanoparticles with uniform diameters (10-15 nm), owing to self-assembly activity of hFTN-H. Each nanoparticle is subjected to carrying a large number of epitopes with a homogeneous and stable conformation per autoantibody binding, thereby allowing substantial enhancement of sensitivity. The sensitivity was finally boosted to 3 attomolar concentration of the autoantibodies, 4-9 orders of magnitude more sensitive than conventional immunoassays. Also, this ultrasensitive protein nanoprobe successfully detected natural autoantibodies in the sera from Type I diabetic patients. The attomolar sensitivity was successfully reproduced upon the detection of other antibodies, i.e. monoclonal antibodies against hepatitis B surface antigen. With the two protein markers above, the feasibility of simultaneous and multiplexing-mode detection was also demonstrated.

489e Direct Photocontrol of the Enzyme Form-Dynamics-Function Relationship

Andrea C. Hamill, Serena Wang, and **C. Ted Lee Jr.**. Department of Chemical Engineering and Materials Science, University of Southern California, 925 Bloom Walk, HED 207, Los Angeles, CA 90089-1211

To properly study protein structure requires both (1) a means to induce changes in the conformational state of proteins, i.e., folding or association, and (2) a means to study the conformation of these intermediately-folded or aggregated structures, preferentially at high resolution. To achieve point #2, work in our lab has revealed that small-angle neutron scattering (SANS) is a powerful technique to study the conformation of partially-folded structures, as well as protein n-mers. Unlike X-ray crystallography, SANS is a solution technique and can provide a more accurate representation of the state of the protein in vivo (e.g., deviations from the native state or self-association equilibrium, etc.) free from crystal packing constraints. Furthermore the often-restrictive need to crystallize the protein is removed (particularly challenging for partially-folded proteins with exposed hydrophobic domains and membrane proteins that inherently contain large hydrophobic moieties).

To achieve point #1, we have developed a novel method to control protein conformation with simple light illumination through the interaction of proteins with photoresponsive surfactants. Photoresponsive surfactants can in essence be switched "on" or "off" with exposure to visible or UV light, respectively, through the photoisomerization of an azobenzene group in the surfactant tail. Hence, surfactant binding to the protein and, thus, protein conformation can be tuned with light.

Through this combined strategy of inducing protein conformational changes with light illumination and studying the high-resolution structure of the resulting conformations with SANS, we have been able to study a variety of protein structural events. For example, we recently discovered a new folding intermediate in lysozyme, where SANS data combined with secondary structural information from FT-IR revealed that unfolding occurs primarily in the α domain of lysozyme, while the β domain remains relatively intact. Thus, this "surfactant-unfolded intermediate" of lysozyme was found to be a distinct structure from the well-known α-domain intermediate, which contains a folded α domain and unfolded β domain. In addition, since the swelling of the α domain was found to leave the active site of the enzyme intact (which could be seen from the SANS-determined structures), the light-initiated transition to the surfactant-unfolded intermediate results in an increase in enzyme flexibility and leads to enzyme "superactivy".

Our combined approach has also revealed that, in addition to protein folding, light illumination can be used to induce changes in protein self-association. Results with α-chymotrypsin demonstrate that under visible light the protein exists as the monomer with the active site exposed (surfactant switched "on", thus, protein-protein interactions are replaced with protein-surfactant interactions), while under UV-light the hexamer form is present, thereby preventing access to the active site. Thus, this novel ability to control enzyme form with light (i.e., quaternary structure) can in turn be used to photo-control enzyme function (i.e., activity). Furthermore, SANS data are used to probe for the first time the high-resolution structure of α-chymotrypsin hexamers in solution. It is found that the protein self-associates through the formation of cork-screw type structures. Due to crystal packing constraints, these structures are undetectable with X-ray crystallography.

489f A Blue Fluorescent Protein with Oxidoreductase Activity

Karen M. Polizzi, School of Biosciences, University of Exeter, Geoffrey Pope Building, Stocker Rd, Exeter, United Kingdom, Desmond A. Moore, School of Physical Sciences and Engineering, Morehouse College, Atlanta, GA 30314, and Andreas S. Bommarius, Georgia Institute of Technology, School of Chemical Engineering, 315 Ferst Drive N.W., Parker H. Petit Biotechnology Institute, Room 3310, Atlanta, GA 30332-0363.

Fluorescent proteins revolutionized the study of proteins, organelles, and metabolic pathways by making it possible to visualize events within cells. In 2001, Su et al identified a blue fluorescent protein from Vibrio vulnificus (Bfpvv) that does not adopt a beta-can fold like traditional fluorescent proteins, but belongs to the short-chain dehydrogenase/reductase family and obtains its fluorescence by binding and magnifying the intrinsic fluorescence of NADPH. Since the fluorescence level of Bfpvv is very low, its natural role is unlikely to be as a fluorescent protein. However, a protein with both fluorescence and activity would be a useful addition to the toolbox for protein biophysical studies as it would allow the separate monitoring of binding and catalytic events.

Despite the diverse nature of the SDR family, we were able to identify redox substrates on which Bfpvv shows considerable activity. The enzyme is highly active in the reduction of aldehydic substrates, particularly those with high hydrophobicity (e.g. aromatic and long chain hydrocarbon). The activity with ketones is considerably lower, as is the activity in the oxidative direction. When oxidizing substrates, the enzyme shows more than 15-fold preference for the S-enantiomer. The substrate profile of the enzyme suggests that the active site is partially impaired and this is what likely led to the development of fluorescent properties. Mutants with increased fluorescence have accumulated mutations in areas that affect the protein activity, further corroborating that fluorescence comes at the expense of activity. The unique properties of Bfpvv allow for both visual levels of fluorescence to monitor binding of NADPH and measurable levels of redox activity to monitor catalytic turnover.

490a Development of Quantum Dot Encoded Silica Beads for Use in An Ultra-Miniaturized Microarray Platform

Gerson R. Aguirre[1], Alexander Couzis[2], Charles Maldarelli[3], Nikhil D. Kalyankar[2], and Lane Gilchrist[2]. (1) Chemistry Department, City College and Graduate Center of the City University of New York, 160 Convent Ave., New York, NY 10031, (2) Chemical Engineering, City College and the Graduate Center of the City University of New York, Steinman Hall, 140th St @ Convent Ave, New York, NY 10031, (3) Chemical Engineering Department and Benjamin Levich Institute, City College and the Graduate Center of the City University of New York, Steinman Hall, 140th St @ Convent Ave, New York, NY 10031

The focus of this presentation is the development of optically barcoded silica beads for use in high-throughput, multiplexed screening applications such as protein microarrays or flow cytometry. Luminescent semiconductor core-shell, CdSe/ZnS QDs nanocrystals (or quantum dots (QDs)) with different emission wavelengths (colors), and encapsulated in different compositions in silica beads are used to define an optical barcode. Quantum dots are usually fabricated so they are stably dispersed in an organic solvent. This is accomplished by terminating the qdot surface (capping) with an tri-alkyl phoshine or a long chain amine. In order to facilitate the incorporation of the qdot into the hydrophilic silica environment it is necessary that the qdots are dispersible in water. We have achieved this first step by coating the qdots with polyvinyl-pyrolidine (PVP). Following this step, the water stable qdots are incorporated in a Stoeber approach that results in the encapsulation of the PVP coated qdots in polymerizing silica. Confocal laser scanning microscopy (CLSM) images of the nanocrystal luminescence in the bead interior indicate that the QDs are segregated into inclusions distributed throughout the bead. CLSM and fluorimetry measurements of the emission spectra of PS beads embedded with three color QDs in varying concentrations are reported which verify that distinguishable optical barcodes derived from the spectral scans of these QD can be obtained by this technique.

490b Optimal Spacing of Zinc Selenide Quantum Dots for Biological Sensing Applications

Bing C. Mei and T. J. Mountziaris. Chemical Engineering, University of Massachusetts, Amherst, MA 01003

Semiconductor nanocrystals or Quantum Dots (QDs) are small crystalline grains of a few hundred to a few thousand atoms with diameters less than 10 nanometers. QDs exhibit size-dependent properties, including size-dependent luminescence, and continuous excitation by any wavelength smaller than the emission wavelength. Their absorption onset and emission peak shift to smaller wavelengths with decreasing size. In addition, QDs have symmetric and narrow emission peaks and their emission spectra do not change with the excitation frequency. Due to their unique properties, QDs have attracted attention as a replacement for organic fluorescent dyes in biosensor applications. Traditional tagging of biomolecules that employs organic fluorophores has significant limitations when compared to tagging using QDs. Organic fluorophores tend to have narrow excitation window, and often exhibit broad emission spectra with red tailing. This makes it difficult to observe different fluorophores using a single excitation source. Furthermore, simultaneous quantitative evaluation of relative amounts of multiple analytes present in the same sample is difficult due to spectral overlap. QDs have high quantum yields, large extinction coefficients, long fluorescence lifetimes, high sensitivity and excellent photostability.

For biological sensing applications, it is important to obtain high quality QDs that exhibit high luminescence emission intensity (i.e. high quantum yield). In addition to the quality of individual QDs, their concentration in a liquid sample plays a critical role in determining the observed intensity of emitted light that reaches the detector. The light propagation in concentrated colloidal suspensions of particles with sizes comparable to the wavelength of the radiation has been a subject of research for several years due to the nonlinear interaction between radiation and particles. In the case of QDs, the size of the particles is much smaller than the wavelength of both the excitation and emitted radiation. Since QDs absorb as well as emit light the behavior is potentially more complex. The absorption of the excitation beam by the particles and the re-absorption of emitted light by slightly larger particles are dependent on the particle concentration along the path of the beam.

In this study we estimate the optimal concentration, and average inter-QD spacing, that maximizes the apparent emission intensity from a liquid sample. This is useful for designing biosensors based on QDs with the purpose of maximizing their detection sensitivity.

Zinc selenide quantum dots (ZnSe QDs) were synthesized using the hot coordinating solvent technique. In a typical synthesis experiment, aliquots were extracted at 10, 20 and 30 minutes of reaction time to produce QDs of increasing size as evident by the red shift in the wavelength of the peak PL intensity with reaction time. The sizes of the three synthesized samples were determined to be 2.7nm, 3.2nm and 3.5nm, respectively. After synthesis, the as-prepared QDs were dispersed in butanol and their emission spectra were measured at various dilutions. Three fluorometers were used to confirm that the observations were not instrument dependent.

The effects of particle size on the optimal QD concentration were studied using a series of ZnSe QD samples of varying average diameters diluted in butanol. Each sample exhibited a maximum PL intensity at a certain QD volume fraction that decreased with particle size. To correct for the variation of particle number density at various dilutions, the peak PL intensity values were divided by the number of QDs in the dispersion and plotted versus QD volume fraction. The normalized PL intensity exhibits a plateau at low QD concentrations, corresponding to the apparent emission intensity from a single QD.

The optimal particle concentration was used to estimate the minimum possible distance between neighboring QDs in dispersion, before attenuation of the PL intensity arises due to high particle concentration. This information is very useful for maximizing the sensitivity of biosensors utilizing QDs as luminescent beacons.

Towards an in Vitro Model of Anti-Therapeutic Resistance: Cellular Drug Efflux Pump Systems in Supported Lipid Bilayer/Nanostructure Hybrid Structures

Bin He, M. Lane Gilchrist
The City College of New York, The Graduate Center, CUNY

Our current studies are focused on the implementation of a biomimetic approach to create solid-supported membranes interfaced with nanostructures. Nanoporous membrane structures have been employed as a new substrate used to functionally immobilize the members of the ATP binding cassette (ABC) transporter superfamily. These assemblies are being used to build highly miniaturized in vitro models of multidrug resistance for high-throughput screening and drug discovery.

The drug efflux pump PDR5-GFP has been isolated from S. cerevisiae clones from a commercial library (InVitrogen) using affinity chromatography. We have constructed anti-GFP Sepharose beads via CNBr coupling with high binding capacities as support for large-scale affinity purification and have isolated PDR5-GFP at 0.1 mg scale. Furthermore, we've also begun to biosynthesize the human multidrug resistance transporter ABCG type 2 using Cell-Free Protein Synthesis (CFPS). Confocal fluorescence microscopy was utilized to analyze supported membranes constructed on 5 micron silica beads and in nanoporous anodic aluminum oxide (AAO) membrane filters with 200 nm and 100 nm pore diameters. We have begun to study the mobility and localization of transporters in supported membrane systems and have initiated fluidity studies of the supported membranes using fluorescence recovery after photobleaching (FRAP). Confinement of transporter assays to nanopores could lead to a highly miniaturized method for the analysis of ABC transporters that utilize miniscule sample amounts such as obtained from tissue biopsies.

Enzyme Functionalized Single Wall Carbon Nanotubes for Biotechnology Use

Pu Zhang, David B. Henthorn

Department of Chemical & Biological Engineering
University of Missouri-Rolla, 143 Schrenk Hall, Rolla, Missouri, 65409

Abstract: Different methods of functionalizing single wall carbon nanotubes with model enzyme were investigated. Enzyme loading and retention of activity were compared among covalent and noncovalent methods. Possible mechanisms of loss of activity were also discussed. In addition, we also introduce a novel photoinitiated "graft from" polymerization method to functionalize polymer molecules on the sidewall of carbon nanotubes. Reaction time was reduced, and the lowered reaction temperature also provides a possible means to attach biomolecules and polymer spontaneously.

1. Introduction

Carbon nanotubes have been recognized as excellent platforms for bioanalytical devices because of their unique properties [1-3]. For example, the conjugation of biomolecules with molecular recognition abilities allows for the construction of sensitive and specific elements. Covalent and non-covalent functionalization methods are widely used for carbon nanotube-biomolecule conjugation. For covalent functionalization, biomolecules are attached via chemical reactions, usually after the nanotube surface has been altered to introduce reactive groups. On the other hand, non-covalent functionalization techniques rely on physical adsorption of the molecule on the CNT surface, mostly through hydrophobic, π-π hydrophobic, or van der Waals forces [4].

Before carbon nanotubes may be effective in the biological or therapeutic fields, the dispersability of the normally hydrophobic nanotube in aqueous solutions must also be addressed. Previous work has focused on solving this problem using either a surface oxidation process, the addition of surfactants, or one of many other techniques [5-7]. In this article, we report a two step "graft from" method where structures rich in polyethylene glycol (PEG) are introduced to the CNT surface. The macromonomer poly(ethylene glycol) monomethacrylate (PEGMA) is used in the surface initiated free radical reaction used to functionalize the carbon nanotube surface. The photoinitiated polymerization allows for lower reaction temperatures compared to traditional thermal or sonically initiated polymerization. The dispersability of the functionalized carbon nanotube was significantly enhanced.

2. Experiments

2.1 Functionalization with model enzyme

To prepare SWNT with covalently attached enzymes, sonochemical treatment was first used to introduce carboxyl groups on the surface. These groups were used as sites to react with surface exposed amine groups on the enzyme. In contrast, noncovalent functionalization was done via simple physical adsorption on pristine CNTs. In both methods, an enzyme solution was prepared and diluted to 2 µg/µl. Forty microliters of enzyme solution were added to 5 ml of a 0.15 mg/ml of SWNT solution for overnight interaction. Unbound enzyme was removed using centrifugation at 15,000x g for 30 minutes where the supernatant was kept for testing enzyme concentration. We used a BCA (bicinchoninic acid) assay to quantify the concentration of AP remains in the supernatant, so the immobilized enzyme could be indirectly quantified. The activity of the immobilized alkaline phosphatase was tested using the substrate PNPP. A PNPP solution (One Step PNPP, Pierce Biotechnology) was added to the SWNT solution and allowed to react in a 96 well microplate. The absorbance at 405 nm was recorded in a microplate reader at various times. The identical procedure was done using the same concentration of unbound, freely dissolved enzyme as a reference.

2.2 Functionalization with PEG-rich polymer grafts

In order to obtain better dispersed carbon nanotubes. HiPCO™ SWNTs (30 mg) were dispersed in 75 ml of a 50:50 (v/v) water, ethanol aolution and sonicated for 2 h at room temperature. The free radical photoinitiator, 1-hydroxycyclohexyl phenyl ketone (HCPK) was introduced at 1% before ultraviolet light (365 nm, 100 mW/cm^2) treatment was carried out with directly irradiation into the mixture in a nitrogen atmosphere. After 30 minutes of UV exposure the reaction was stopped. Excess unbound HCPK molecules were removed by centrifugation, washing, and resuspending. PEGMA monomer (2 g) was then introduced and polymerization was done with another 30 minute dose of UV irradiation. Free monomer was removed by repeated vigorous centrifugation and filtration steps.

3. Results and discussion

3.1 Enzyme functionalized nanotubes

After sonochemical treatment, the SWNT samples were incubated in enzyme solutions buffered at either a pH of 4 or 7. The enzyme loading efficiency was evaluated for both, but was extremely low for the pH 7 samples. Sample incubated at a pH of 4 were found to have 45 µg of enzyme per mg of SWNT, according to the BCA assay. In contrast, the loading efficiency of SWNT-enzyme conjugate formed by physical adsorption was higher at 66 µg/mg. To compare retention of enzyme activity, reaction with the substrate PNPP was carried out. For sonochemical functionalization, only 10% enzyme activity remained when compared to an equivalent amount of enzyme in free solution. Functionalization using physical adsorption yielded significantly better enzyme loading and retention of activity, with the functionalized nanotubes retaining 58% of the enzyme activity. Figures 1 (a) and 1 (b) show the measurements of enzyme activitiy on sonochemical treated SWNTs and those functionalized by physical adsorption, respectively. The loss of enzyme activity could be attributed to the blocking of active sites since the physical adsorption process produces random orientations.

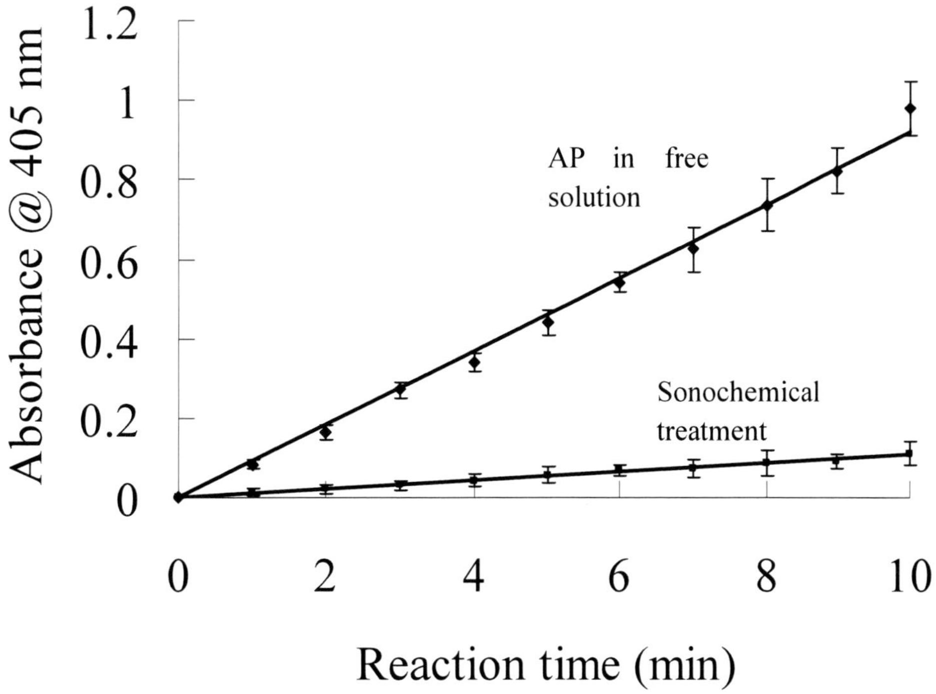

Figure 1 (a) Measurement of AP activity on sonochemically treated SWNTs

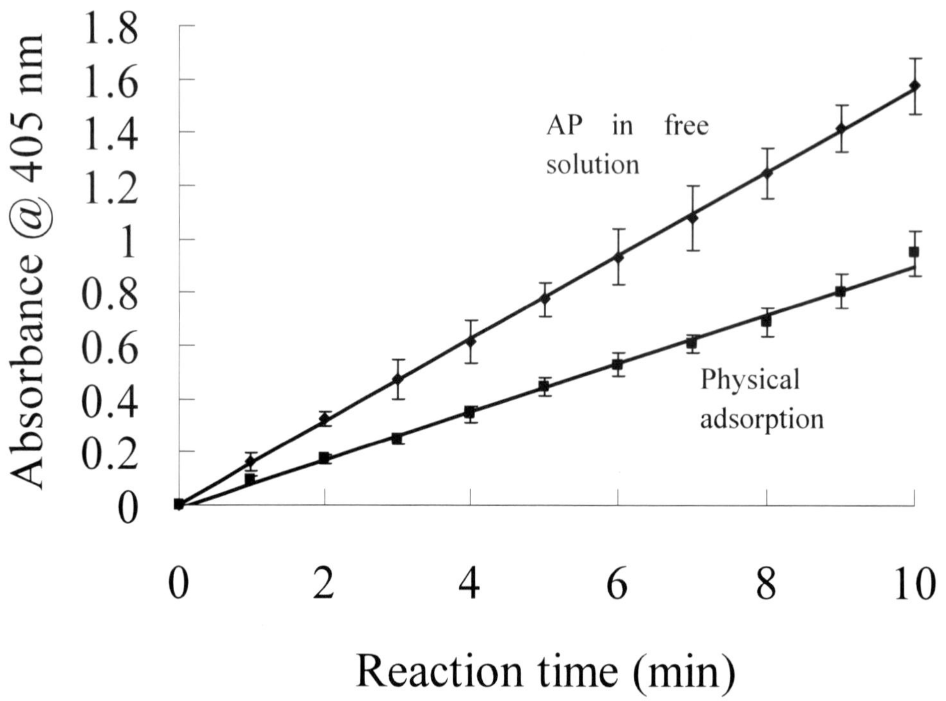

Figure 1 (b) Measurement of enzyme activity after physical adsorption on SWNTs

3.2 Grafted PEG-rich structures on SWNTs

We used attenuated total reflectance FTIR to characterize the polymer functionalized carbon nanotubes obtained after different polymerization times. The FTIR spectrum of the pristine carbon nanotubes (Figure 2a) is featureless. The peak at ~2869 cm^{-1} corresponding to the stretching and vibration of C-H bond which is barely observed in 5 min polymerization sample and but is clear in the 15 min and 30 min sample. The peaks at ~1100 cm^{-1} were attributed to the PEG group of the polymer.

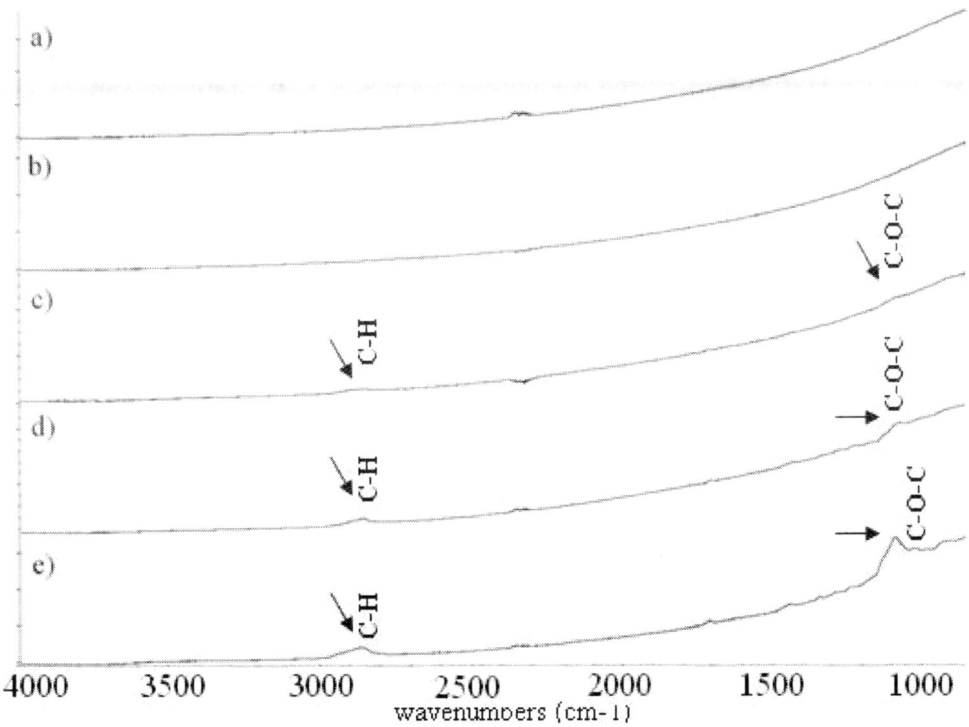

Figure 2. FTIR spectra of functionalized carbon nanotubes a) Pristine nanotubes,, b) Nanotubes polymerized in the presence of PEGMA after 1 minute, c) 5 minutes, d) 15 minutes, and e) 30 minutes of polymerization time.

A control experiment was also done by simply mixing carbon nanotubes and PEGMA but no UV exposure was carried out. Following the same purification procedures, no PEG was seen grafted on the CNTs.

Resuspension of the functionalized carbon nanotubes was easily carried out, and the dispersability of the modified carbon nanotube was enhanced as can be seen in Figure 3.

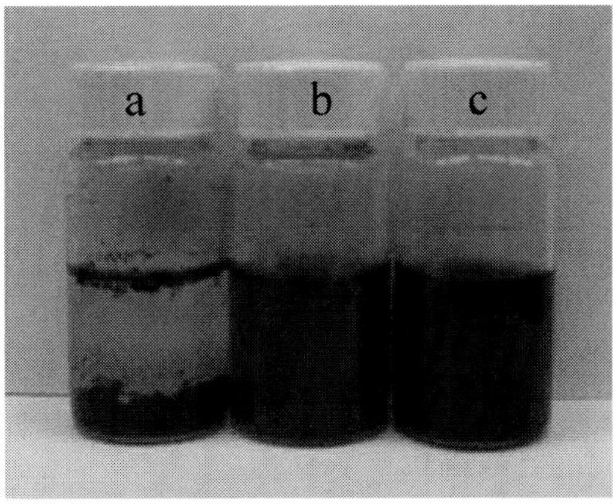

Figure 3. Photographs of SWNTs dispersed in water. a) Pristine SWNTs, b) SWNTs with grafted PEG-containing polymers after 15 min polymerization, and c) after 30 min of polymerization.

4. Conclusions

In this work, we evaluated both covalent and noncovalent methods for the immobilization of a model enzyme on carbon nanotube surfaces. The simplest method, physical adsorption of enzyme on unmodified CNT surface, yielded the best results. The alkaline phosphatase molecule was added at a loading of 66 μg enzyme per mg of nanotube. At this loading, the enzyme molecules retained 58% of their activity. In addition, a novel photoinitiated "graft from" polymerization method to produce polymer functionalized carbon nanotube is also introduced. The attached polymer molecules significantly enhance aqueous dispersability, and the low reaction temperature creates a new environment for making polymer coated carbon nanotube-biomolecules conjugates.

References

[1] M. Dresselhaus, M. Endo. Carbon Nanotubes Synthesis, Structure, Properties, and Application. Spinger. 2000

[2] L. Dai, A. W. H. Mau, Surface and Interface Control of Polymeric Biomaterials, Conjugated Polymers, and Carbon Nanotubes. J. Phys. Chem. B 2000, 104, 1891-1915

[3] M. Hughes, E. Brandin, J. Golovchenko. Optical Absorption of DNA-Carbon Nanotube Structures.

[4] T. Elkin, X. Jiang, S. Taylor, Y. Lin, L. Gu, H. Yang, J. Brown, S. Collins, Y. Sun. Immuno-Carbon Nanotubes and Recognition of Pathogens. ChemBioChem, 2005, 6, 640-643

[5] J. Stevens, A. Huang, H. Peng, I. Chiang, V. Khabashesku, J. Margrave. Sidewall Amino-Functionalization of Sinlge-Walled Carbon Nanotubes through Fluorination and Subsequent Reactions with Terminal Diamines. Nano Letters, 2003, 3 (3), 331-336.

[6] B. Zhao, H. Hu, A. Yu, D. Perea, R. Haddon. Synthesis and Characterization of Water Soluble Single-Walled Carbon Nanotube Graft Copolymers. J. AM. CHEM. SOC. 2005. 127, 8197-8203.

[7] D. Hill, Y. Lin, A. Rao, L. Allard, Y. Sun. Functionalization of Carbon Nanotubes with Polystyrene. Macromolecules 2002, 35, 9466-9471

490f Probing Biological Samples Using Cryo-Tem

Ashish Jha and Arijit Bose. Chemical Engineering, University of Rhode Island, Kingston, RI 02881

Cryogenic transmission electron microscopy (cryo-TEM) has been used to probe biological samples such as proteins, human-saliva and human-serum. Protein images show detailed features such as size and morphology of individual BSA (bovine serum albumin) molecule, emphasizing its oblate ellipsoid (14 nm x 4 nm) structure. Human-saliva being a non-invasive biological fluid is an attractive alternative of serum for many purposes such as biomarker studies. Cryo-TEM reveals a variety of structures in whole human saliva (WHS), which includes globular mcelles (~5nm), vesicles (~100nm) and huge clusters of micelles (~100-500nm). Images of human-serum have features ranging from globular proteins and circular-discs to highly polydisperse unilamellar vesicles and multilamellar vesicles. One special feature that has been observed in serum is in the shape of a semicircular arc joined with a circular object. Cryo-TEM has been found to provide enough details to be a viable tool for studying biological samples at nanoscale resolution and may be potentially useful in drug efficacy testing and biomarker applications.

Development of siRNA Nanoparticles for Targeted Delivery

Gunilla B. Jacobson-Andrews[1], *Rajesh R. Shinde*[2], *Robyn P. Hickerson*[3], *Roger L. Kaspar*[3], *Christopher H. Contag*[2] *and Richard N. Zare*[2]

[1]Department of Chemistry, Stanford University, Stanford, CA.
[2] Departments. of Radiology, Microbiology & Immunology, Pediatrics, and Pathology, Stanford University School of Medicine, Stanford, CA
[3]TransDerm Inc., Santa Cruz, CA

RNA interference (RNAi) is rapidly gaining importance as a set of tools to control gene expression and has the potential to develop into new classes of drugs for therapy. The full potential of these new compounds has been stunted by the lack of appropriate delivery mechanisms. In this project, we have developed nanoparticles of small inhibitory RNA (siRNA) using the biodegradable polymer poly(lactic acid) (PLA) and the copolymer PLA/poly(ethylene glycol) (PLA/PEG). Supercritical carbon dioxide (SC-CO_2) was used as an anti-solvent and the drug/polymer solutions were sprayed into the supercritical fluid (SCF) to form particles by rapid precipitation. The high compressibility of SCF allows for tunable solvent properties and tight process control that enable optimization of particle size and particle size distribution. Due to the unique properties of SC-CO_2, no purification or drying steps are needed, and the process is compatible with bioactive compounds, such as proteins and nucleic acids.

Initially, unmodified siRNA targeting the mRNA encoding mutant N171K keratin 6a (K6a: this mutation results in the skin disorder pachyonychia congenita) was processed using the SCF process yielding siRNA nanoparticles in the sub-100 nm range. The resulting siRNA particles were analyzed by scanning electron microscopy and dynamic light scattering for particle size distribution. A functional cell culture assay was performed to ensure biological activity of the siRNA after particle formation. In this assay the K6a cDNA was fused to a luciferase reporter, resulting in a bicistronic expression vector that allowed rapid detection of the reporter construct and activity of the siRNA nanoparticles. The siRNA mediated reduction of the keratin producing gene was correlated with light emission in live cell assays, and this revealed that siRNA activity was preserved after the SCF process.

Subsequently, the siRNA particles were encapsulated using PLA or PLA/PEG. The release of active siRNA was shown to be highly dependent on polymer concentration, polymer molecular weight, and SCF process conditions (such as flow rate, temperature and pressure). Encapsulated siRNA particles formed particles in the 200-500 nm range which could be controlled by process conditions. SCF processes offer new approaches for the formation of biologically active nanoparticles that may lead to effective tools for controlled delivery of siRNA

545a Multi-Component Nanoparticles for Combined Fluorescence, Optical and Magnetic Resonance Imaging

Marian E. Gindy and Robert K. Prud'homme. Chemical Engineering, Princeton Unversity, E-Quad, Olden Street, Princeton, NJ 08544

Recent advances in synthesis strategies for the fabrication of nanoscale contrast agents has lead to significant advancements in the understanding of biological processes at the molecular level, the development of better diagnostic tools, and the ability to accurately monitor drug uptake for improved drug dosing. Nanocrystalline materials such as fluorescent-doped silica nanoparticles, quantum dots and paramagnetic nanomaterials have overcome many of the early limitations associated with conventional organic dye contrast agents, including poor photostability, low quantum yield and insufficient biological stability. However, challenges in the controlled and reproducible synthesis of biocompatible, bioinert nanomaterial contrast agents remain unmet.

In this work, we demonstrate the successful preparation hybrid nanoparticles for multi-modal imaging applications. Block copolymer nanoparticles successfully encapsulating organic fluorescent dyes, metal nanostructures and superparamagnetic materials are prepared for use in combined fluorescence, optical and magnetic resonance imaging applications. The nanoparticle preparation process, termed Flash NanoPrecipitation, relies on the controlled self assembly of incorporated agents and subsequent stabilization by a biocompatible amphiphilic block copolymer to yield highly stable nanoparticles with controlled size (50-300 nm), narrow particle size distributions, specific component composition, and high loading efficiencies. Because the process herein described relies solely on the control of mixing and aggregation timescales, and not on intrinsic physical properties of imaging components, it is anticipated that it can be expanded to include a wide variety of novel contrast agents, yielding nanoparticle formulations for integrated drug delivery and imaging applications.

545b Carbon Nanomaterials for Medical Imaging

Pratixa P. Joshi, Valerie C. Moore, S. Ward Casscells, and Jodie L. Conyers. Office of Biotechnology, University of Texas Health Science Center at Houston, 6770 Bertner Ave,, Room c954, Houston, TX 77030

Carbon nanomaterials possess great potential to be used in variety of fields. Our group focuses on using various carbon nanomaterials to enhance imaging properties of a variety of MRI and CT-based contrast agents. The present study demonstrates that carbon nanomaterials, including single-walled carbon nanotubes, double-walled carbon nanotubes, multi-walled carbon nanotubes and graphite, associate with the imaging active agents and enhance their ability for medical imaging. For CT imaging agents we have focused on iodine and for MRI agents gadolinium, manganese, erbium and chromium. In addition to the synthesis, physical characterization, and imaging properties, the results from the in-vitro analysis of the different materials will be presented.

545c Nano-Metal Particles for Fluorescence Enhancement in Fluorophore Mediated Biosensing and Bio-Imaging

Kyung A. Kang, Jianting Wang, Bin Hong, and Hanzhu Jin. Chemical Engineering, University of Louisville, Ernst Hall 106, Speed School of Engineering, University of Louisville, Louisville, KY 40292

Nanometal particles have high-density surface plasmon polariton fields, and when a fluorophore is placed within these fields, the lone pair electrons of the fluorophore, which normally participate in the fluorescence self-quenching, are transferred to the surface plasmon field, resulting in artificial fluorescence enhancement. Therefore, nanometal particles can be used to improve the effectiveness of fluorophore mediated biosensor and bio imaging by amplifying the fluorescence signals from the fluorophore at target locations. We have tested the fluorescence enhancement by nano-gold particles both in optic-fiber based biosensing system and in molecular imaging of breast cancer. The level of enhancement was found to depend upon the metal type, size of the particle, quantum yield of a fluorophore, and, most importantly, the distance between the nanoparticle and the fluorophore. Various ways to control the distance between a fluorophore and an NMP are applied, depending on the application. For fiber-optic immuno-sensing, as an in-vitro application, a surfactant layer was added at a particular thickness on the NMP before mixed with fluorophores, and the concentrations of the fluorophores and NMPs in solution are adjusted. For tumor model imaging, as an in-vivo study, fluorophores were covalently linked onto NMPs via a spacer at an appropriate size to ensure these two entities were placed together at the tumor site. For bio-sensing system, the fluorescence signal from the fluorephore was enhanced by as much as 200% with nano gold particles. In bio-imaging, the fluorescence of nanoparticle-linked Indocyanine green was enhanced by 800 times. Authors acknowledge the partially financial support from National Science Foundation (BES-0330075) on biosensing study. The authors also thank the U.S. Army Medical Research and Materiel Command (DAMD17-03-1-0572) for partial financial support. The authors also Dr. Achilefu's group at Washington University, MO for supplying Cypate.

545d Biocompatible Silicon Quantum Dots

Folarin Erogbogbo[1], Ken-Tye Yong[2], Paras N. Prasad[2], and Mark T. Swihart[1]. (1) Chemical and Biological Engineering, The University at Buffalo (SUNY), 303 Furnas Hall, Buffalo, NY 14260-4200, (2) Institute for Lasers, Photonics, and Biophotonics, The University at Buffalo (SUNY), Buffalo, NY 14260-4200

Silicon quantum dots (QDs) have great potential for use in biological imaging and diagnostic applications, as an alternative to fluorescent organic dyes or potentially toxic heavy metal containing quantum dots. For quantum dots to be useful in biological systems they must remain stably dispersed in water and biological fluids, over a wide range of pH and salt concentration. There have been many challenges in creating such stable, water-dispersible quantum dots, including instability of photoluminescence due to oxidation in an aqueous environment and difficulty in attaching hydrophilic molecules to silicon surfaces without inducing oxidation. Here, a highly stable aqueous suspension of silicon QDs has been prepared, that maintains all of the desirable optical properties of Si nanocrystals dispersed in nonpolar solvents. These QDs are prepared by first grafting styrene to the surface of bare silicon nanocrystals, followed by suspension in a chloroform solution containing phospholipids. The solvents are evaporated and the particles are then dispersed in water, prompting the encapsulation of the silicon nanocrystals by phospholipid micelles. Phospholipids with different functional groups including carboxylic acid, biotin, and amine groups can be used in this process, providing built-in surface functionalization. This will allow silicon quantum dots to be useful for a variety of biological applications. The QD stability and optical properties are characterized in biologically relevant environments.

545e Development of Nanoparticles for in Vivo Imaging

Gunilla B. Jacobson-Andrews[1], Rajesh Shinde[2], Christopher H. Contag[2], and Richard N. Zare[1]. (1) Department of Chemistry, Stanford University, 333 Campus Drive #121, Stanford, CA 94305, (2) Departments of Pediatrics, Radiology and Microbiology and Immunology, Stanford University, 333 Campus Drive #121, Stanford, CA 94305

Nanotechnology is gaining importance in biomedical applications for drug delivery and imaging. It has the potential to enhance the efficacy and specificity of drugs and imaging agents and allow for a more detailed understanding of biological systems. For e.g., in cancer therapy the challenges include early disease detection and targeted delivery of imaging and therapeutic agents to the tumor region. Most cancer drugs do not discriminate between normal and diseased tissues causing undesirable side effects. A potential approach to improved therapeutic efficacy of these drugs is better drug delivery system that can target tumors with high specificity. Non-invasive imaging modalities offer a potential framework for developing such drug delivery systems and can provide spatio-temporal information of the delivery and release of model imaging agents at diseased sites. In this study, we use BLI to develop novel drug delivery systems based on encapsulated luciferin nanoparticles, with luciferin as the model drug. Nanoparticles of luciferin have been formed using a supercritical fluid antisolvent system (SAS), and process conditions have been optimized for particle size and particle size distribution. Due to the unique properties of supercritical carbon dioxide, no purification or drying steps are needed, and the process is compatible with bioactive compounds, such as proteins. The particles have also been encapsulated, using the SAS process with biodegradable polymers such, as poly(lactic acid) (PLA) and poly(DL-lactide-co-glycolide) (PLGA), for controlled-release of drug and to protect active compounds from degradation during storage and delivery. A transgenic reporter mouse that expressed luciferase ubiquitously throughout the mouse is used to study the biodistribution of the model drug, luciferin. Luciferin is delivered as a modified nanoparticle either by inhalation or intravenously, and its biodistribution imaged noninvasively.

545f Monodisperse Microbubble Contrast Agents for Improved Ultrasound Contrast Imaging

Esra Talu[1], Kanaka Hettiarachchi[2], Shukui Zhao[3], Robert L. Powell[1], Abraham P. Lee[2], Marjorie L. Longo[1], and Paul Dayton[3]. (1) Chemical Engineering and Materials Science, University of California, Davis, Davis, CA 95616, (2) Biomedical Engineering, University of California, Irvine, Irvine, 92697, (3) Biomedical Engineering, University of California, Davis, Davis, CA 95616

Encapsulated microbubbles are used as contrast agents for molecular imaging with ultrasound. Current techniques of producing microbubbles result in a large size variance. Detection of these contrast agents depends on properties related to the microbubble size such as resonant frequency, and current ultrasound imaging systems have bandwidth limits which reduce their sensitivity to a polydisperse contrast agent population. For ultrasonic molecular imaging, where only a limited number of targeted contrast agents may be retained at the site of pathology, it is important to optimize sensitivity of the imaging system to the entire population of contrast agent. In this manuscript, a microfluidics technique was used to produce monodisperse microbubbles having a narrow size distribution specifically for ultrasound molecular imaging. Monodisperse microbubbles within the diameter range of 1-10 microns can be made with a diameter standard deviation on the order of 0.2 microns. These results are in contrast to currently available lipid shelled agents, which we measure to have a standard deviation in diameter on the order of 1 micron. We demonstrate the manufacture of monodisperse functionalized and targeted lipid-shelled microbubble contrast agents for molecular imaging applications, and we demonstrate acoustically that our monodisperse agents demonstrate a higher pulse to pulse echo correlation compared to that of polydisperse microbubbles (0.9 vs. 0.7). Our initial results indicate that optimizing contrast agent distribution has the potential to enhance the sensitivity of contrast imaging.

558a Single Walled Carbon Nanotubes as Single Molecule Chemical Sensors within Living Cells

Michael S. Strano[1], Hong Jin[2], Daniel A. Heller[3], and Jong Hyun Choi[1]. (1) Department of Chemical and Biomolecular Engineering, University of Illinois at Urbana-Champaign, 600 S. Mathews Ave., Urbana, IL 61801, (2) Chemical and Biomolecular Engineering, University of Illinois--Urbana Champaign, box c-3, ral, mc 721, 600 s mathews, Urbana, IL 61801, (3) Department of Chemistry, University of Illinois at Urbana-Champaign, 104 RAL, Box 93-5, MC-712, 600 S. Mathews Ave., Urbana, IL 61801

Chemical agents which damage or bind to nucleic acids are under continuous study for prediction, diagnosis, and treatment of cancer, Alzheimer's disease, and the effects of aging.1,2 Their activity is difficult to measure in real time and in vivo, however, preventing many types of studies on chemotherapeutic agent activity or oxidative stress.3,4 We demonstrate that single-walled carbon nanotubes (SWNT), hollow graphene cylinders with diameters between 0.5 and 2 nm, optically transduce subtle conformational changes in synthetic DNA via spectral changes in near-infrared fluorescence.5,6 We show that a pair of carbon nanotubes can uniquely detect and identify agents of DNA damage in real time within live mammalian cells. Nucleic acid encapsulated nanotube near-infrared photoluminescence is modified upon chemotherapeutic alkylation and reactive oxygen species (ROS) activity via spectral shifts of up to 30 meV. Singlet oxygen, the hydroxyl radical, and alkylating drugs such as melphalan and cisplatin induce differentiating spectral changes between two carbon nanotube species which elucidates the agent involved. These differences are detected from within live mammalian cells. This finding demonstrates the first multiplexed optical detection from a nanoscale sensor. The work provides the first tool to measure alkylating and ROS activity in real-time in living tissue due to the unique photostability and non-invasive near-infrared emission of carbon nanotubes.

This platform has several advantages for fundamental studies of the cellular uptake of nanomaterials. Single-walled carbon nanotubes (SWNT) fluoresce in the near-infrared (n-IR) and are one of a very few classes of fluorophores with no observable photobleaching when excited at moderate fluence. We use this property to track over 10,288 individual trajectories as particles are incorporated into and expelled from NIH-3T3 cells over a period of 127 min on a perfusion microscope stage. A total of 5223 particle trajectories (49.2%) exhibit purely convective diffusion without cellular interaction in the applied flow field. The remaining 5065 trajectories (50.8%) show distinct processes of cellular membrane adsorption (6.2%), surface diffusion (18.4%), endocytosis (12.7%), exocytosis (5.9%) or membrane desorption (7.4%). An analysis of each mean square displacement allows for the first time the complete construction of the network of pathways experienced by nanoparticles as they are trafficked into and within the cell. We observe the first conclusive evidence of nanoparticle exocytosis in this system, and show that the rate closely matches the endocytosis rate with negligible temporal offset. We identify and study a unique pathway that leads to the previously observed aggregation and accumulation of SWNT within the cells. The results have significant implications for the use of nanoparticles in biological systems.

1. Shi, H. L., Hudson, L. G. & Liu, K. J. Oxidative stress and apoptosis in metal ion-induced carcinogenesis. Free Radical Biology and Medicine 37, 582-593 (2004).

2. Povirk, L. F. & Shuker, D. E. DNA-Damage and Mutagenesis Induced by Nitrogen Mustards. Mutation Research-Reviews in Genetic Toxicology 318, 205-226 (1994).

3. Berliner, L. J., Khramtsov, V., Fujii, H. & Clanton, T. L. Unique in vivo applications of spin traps. Free Radical Biology and Medicine 30, 489-499 (2001).

4. Brandes, R. P. & Janiszewski, M. Direct detection of reactive oxygen species ex vivo. Kidney International 67, 1662-1664 (2005).

5. Saito, R., Dresselhaus, G. & Dresselhaus, M. S. Physical Properties of Carbon Nanotubes (Imperial College Press, London, 1998).

6. Heller, D. A. et al. Optical detection of DNA conformational polymorphism on single-walled carbon nanotubes. Science 311, 508-511 (2006).

558b Genetically-Encoded Biosensors of Cellular Metabolites in Plants

Karen M. Polizzi, Takahiro Ishikawa, Nick Smirnoff, and John Love. School of Biosciences, University of Exeter, Geoffrey Pope Building, Stocker Rd, Exeter, United Kingdom

The discovery of fluorescent proteins, such as GFP or BFP, gave researchers the ability to visualize organelles and proteins within cells. Genetically-encoded biosensors, some based on GFP-like proteins, allow the non-invasive, non-destructive monitoring of cellular metabolites at different points in the cell cycle. They offer the opportunity to better understand cell signaling pathways and to monitor in vivo responses to various external forces. Additionally, since sensing occurs upon binding of a ligand, the sensitivity and specificity of sensors can be modulated via mutation.

This talk will focus on two types of protein-based biosensors. FRET-based sensors are a three protein fusion with two fluorescent proteins flanking a ligand binding domain. Upon binding of the ligand, a confirmational change in the receptor region brings the fluorescent proteins into the appropriate configuration for energy transfer, leading to a signal. Alternatively, a binding region can be directly inserted into a single fluorescent protein, as is the case with cameleon GFP calcium sensors. We have explored the use of periplasmic binding proteins, such as maltose binding protein, estrogen receptors, and the auxin binding protein in both FRET and cameleon designs. The advantages and disadvantages of each type of protein-based sensor will be discussed in the context of monitoring metabolite concentrations in living plants.

Drug delivery into the human brain with Diffusion Tensor Imaging

Mahadevabharath Somayaji, Monil Shah, Libin Zhang, Michalis Xenos, Andreas Linninger
Laboratory for Product and Process Design, Depts of Chemical Engineering and Bioengineering,
University of Illinois at Chicago, Chicago, IL 60607 E-mail: {msomay2, linninge}@uic.edu

---

Conference: AICHE Annual Meeting, Salt lake city, UT, USA
Date: Nov 4-9, 2007

Abstract

Motivation: The development of systematic drug delivery for brain disorders is a formidable challenge, and there is hardly any successful treatment protocol for the majority of neurodegenerative diseases. Data published by National Institutes of Health suggest that neurodegenerative diseases such as Parkinson's, Alzheimer's, and Huntington's disease affect millions of people worldwide. Targeted delivery of large macromolecules to specific locations in the brain is a challenging task in the treatment of neurodegenerative diseases of the Central Nervous System (CNS) owing to the presence of blood-brain barrier (BBB) (Pardridge, 1999). In addition, the restriction offered by fibers of the white matter hampers the transport and metabolism of drugs at target areas and pose difficulty to catheter design and placement.

Methodology: In this presentation, we propose three innovations embedded into a hierarchical design procedure. The systematic approach to designing drug targeting faces three challenges and we aim to address them in this presentation:

1. Accurate three-dimensional reconstruction of the *patient-specific* brain geometry
2. Prediction of drug distribution in *anisotropic heterogeneous* brain tissue calibrated from in-vivo patient-specific diffusion tensor data
3. Quantification of *infusate backflow* in high flow micro-infusion regime

The primary task is to accurately reconstruct the physiologically consistent three-dimensional *patient-specific* brain geometry. State of the art geometric image reconstruction tools are used to render and construct computational grids diffusion tensor magnetic resonance imaging data (DT-MRI) (Somayaji et al., 2007).

The second task is to predict distribution of therapeutic agents by accounting for anatomically consistent drug and tissue properties. With sophisticated calibration techniques, the *anisotropic* and *heterogeneous* effective diffusion and hydraulic perfusion tensors are calibrated and used in the first principles model (Tuch et al., 2001; Lazar et al., 2003). In addition, different catheter design such as single and multiple hole catheters are investigated based on volume of drug distribution. The computational model uses clinically relevant catheter designs and accounts for the fluid dynamics inside the catheter lumen. Proper boundary conditions are used to predict the distribution of high molecular weight trophic factor, glial derived neurotrophic factor (GDNF) infused into the Putamen (Somayaji et al., 2007).

In high flow micro-infusion, *backflow* is a frequently occurring physical phenomenon and depends on the catheter design and infusion parameters. The first principles model accounts for the backflow mechanics and tissue deformation for a given choice of infusion and catheter design parameters in addition to considering anatomically consistent brain properties (Chen et al., 1999; Morrison et al., 1999). Development of such a three dimensional holistic model applicable to a human brain will complement clinical trials.

References

Chen M.Y., Lonser R.R., Morrison P.F., Governale L.S., Oldfield E.H., Variables affecting convection- enhanced delivery to the striatum: a systematic examination of rate of infusion, cannula size, infusate concentration, and tissue-cannula sealing time. J. Neurosurg. 90(2), pp. 315-320, 1999.

Lazar, M., Weinstein, D.M., Tsuruda, J.S., Hasan, K.M., Arfanakis, K., Meyerand, M.E., Badie, B., Rowley, H.A., Haughton, V., Field, A., Alexander, A.L. White Matter Tractography using Diffusion Tensor Deflection. Human Brain Mapping. 18(4), pp. 306-321, 2003.

Morrison, P.F., Chen, M.Y., Chadwick, R.S., Lonser, R.R., Oldfield, E.H., Focal delivery during direct infusion to brain: role of flow rate, catheter diameter, and tissue mechanics. Am. J. Physiol. Regul. Integr. Comp. Physiol. 277(4), R1218- R1229, 1999.

Pardridge, W.M. Non-invasive drug delivery to the human brain using endogenous blood-brain barrier transport systems. Pharm. Sci. Technol. Today. 2(2), pp. 49-59, 1999.

Somayaji, M.R; Xenos, M; Zhang, L, Mekarski, M and Linninger A.A., "Systematic Design of Drug Delivery Therapies", *Computers and Chemical Engineering*, In Press, To appear Dec-2007

Tuch, D.S, Wedeen, V.J., Dale, A.M., George, J.S., Belliveau, J.W. Conductivity tensor mapping of the human brain using diffusion tensor MRI. Proc. Natl. Acad. Sci. USA, 98(20), pp. 11697-11701, 2001.

558d Cest Liposomes Provide Multi-Color MRI Contrast for Cell Imaging

Michael T. McMahon[1], Yah-el Har-el[2], Assaf A. Gilad[2], George Sgouros[2], Jeff W.M. Bulte[2], and Peter C.M. van Zijl[1]. (1) F.M. Kirby Research Center for Functional Brain Imaging, Kennedy Krieger Institute, Baltimore, MD 21205, (2) Russell H. Morgan Dept. of Radiology and Radiological Science, Johns Hopkins University, Baltimore, MD 21205

Current cellular MR imaging relies on metallic contrast agents to produce hypo- or hyper-intense spots where cells are located. We present a series of protein-containing liposomes for cellular MRI. These proteins can be selectively activated by frequency-specific RF irradiation where the contrast produced relies on chemical exchange saturation transfer (CEST) from protein amide protons to water. The proteins can be distinguished by the saturation frequency dependence of their particular protein content, also referred to as ‘colors'. By incorporating CEST agents into liposomes, we can limit their interactions with biological media and avoid local dilution of agent. CEST liposomes are capable of producing significant contrast at nM concentrations of liposome. A variety of poly-peptides capable of producing a CEST effect were encapsulated within liposomes, including Poly-L-Ornithine (PLO), Poly-L-Arginine (PLA), and Poly-L-Lysine (PLL). The liposomes display similar frequency dependence of MR contrast to free PLO, PLA, and PLL. Next, it was determined whether the liposomes would enter cells and maintain cell viability. 9L rat glioma cells were incubated overnight in 96 well plates with a series of liposome concentrations, and an MTT assay was used to assess the cell viability. For this level of confluent cells, concentrations of 0.32 nM PLL liposome (18mg/mL PLL solution encapsulated in liposomes) in the media were tolerated, producing viable cells. This new agent, which is composed of purely natural substances should prove useful for future cell tracking studies, and potentially as a diagnostic agent as well.

558e Real-Time Single-Molecule Tracking on the Uptake and Transport Pathway of Single-Walled Carbon Nanotubes in NIH3T3 Murine Cells

Hong Jin, Chemical and Biomolecular Engineering, University of Illinois--Urbana Champaign, box c-3, ral, mc 721, 600 s mathews, Urbana, IL 61801 and Michael S. Strano, Chemical Engineering, Massachusetts Institute of Technology, Building 66, Cambridge, MA 02139.

Single-walled carbon nanotubes(SWNT) fluoresce in the near-infrared(n-IR) and are one of a very few classes of fluorophores with no observable photobleaching when excited at moderate fluence. We use this property to track over 10,000 individual trajectories as SWNT are incorporated into and expelled from NIH-3T3 cells over a period of up to 340min on a perfusion microscope stage. In a typical experiment, a total of 5065 trajectories(49.2%) exhibit purely convective diffusion without cellular interaction in the applied flow field. The remaining 5223 trajectories(50.8%) show distinct processes of membrane adsorption(6.2%), surface diffusion(18.4%), endocytosis(12.7%), exocytosis(5.9%) or membrane desorption(7.4%). An analysis of each mean square displacement allows, for the first time, the complete construction of the mechanistic steps experienced by SWNT as they are trafficked into and within the cell. We observe the first conclusive evidence of SWNT exocytosis in this system, and show that the rate closely matches the endocytosis rate with negligible temporal offset. Endocytosis and exocytosis together explain the previously observed aggregation and accumulation of SWNT within several mammalian cell systems.

606a Self-Assembling Bioactive Protein-Based Hydrogels With Tunable Structural Properties

Ian R. Wheeldon[1], Scott Calabrese Barton[2], and Scott Banta[1]. (1) Chemical Engineering, Columbia University, 500 West 120th Street, New York, NY 10027, (2) Chemical Engineering and Materials Science, Michigan State University, East Lansing, MI 48824-1226

Protein engineering provides an opportunity to rationally combine self-assembly motifs with functional domains to create functionally active supramolecular structures. To this end, we have created self-assembling bioactive hydrogels through the fusion of previously developed hydrogel forming domains to functional proteins. Examples of monomeric (green fluorescent protein, GFP) and tetrameric (red fluorescent protein, DSRED) bioactive-hydrogel forming proteins are presented. Physical cross-linking of the protein monomers occurs due to the agglomeration of the alpha-helical hydrogel-forming domains into coiled-coils and also due to multimer formation. Small amplitude oscillatory shear experiments show that incorporation of the bioactive proteins into the hydrogel structure results in an increase in elastic character as judged by the storage modulus. Additionally, erosion rate is significantly suppressed as the incorporation of bioactive proteins impedes the formation of closed loops. Confocal microscopy and Förster Resonance Energy Transfer (FRET) experiments show that neat bioactive hydrogels and mixtures of the bioactive monomers and a previously designed protein hydrogel monomer are macrohomogenous. The compatibility of the hydrogel forming monomers also allows for the control of micro-scale structure, rheological properties and bulk bioactivity. This protein-engineering based approach to hydrogel design could prove to be a powerful means for constructing enzymatic and stimulus-responsive protein-based hydrogels for active scaffolds in tissue engineering applications and for surface modifications in bioreactor, biosensor and biofuel cell applications.

606b Thermodynamic Peptosome-Based Supramolecular Structures

Jeng-Shiung Jan[1], Jeffery Gaspard[2], James Silas[2], and Daniel F. Shantz[3]. (1) Chemical and Biomolecular Engineering, Georgia Institute of Technology, 311 Ferst Drive, Atlanta, GA 30332-0100, (2) Chemical Engineering, Texas A&M University, 3122 Chemical Engineering, College Station, TX 77843-3122, (3) Department of Chemical Engineering, Texas A&M University, TAMU 3122, College Station, TX 77843-3122

There have been significant advances in the last ten years in the synthesis of "hybrid" block copolymers wherein one block is a poly-á-amino acid and the other a "conventional" polymer (e.g. polystyrene, polybutadiene, PEG) as well as block copolypeptides. These materials have been shown to self assemble into structures including vesicles, micelles, and hydrogels. While these materials are of great interest, they tend to form highly irreversible, i.e. kinetically trapped, structures. Here we report on the self-organization of poly(L-lysine)-b-polyglycine diblock and triblock copolypeptides in aqueous solution. Polyglycine was chosen as the hydrophobic block in the current work in lieu of other hydrophobic amino acids as polyglycine will not form secondary structures such as helices and sheets and has a higher solubility than most other hydrophobic amino acids. These polymers form vesicular structures on the 100-300 nm size scale in solution that appear thermodynamic in nature. A battery of methods including light scattering, conductivity, and microscopy are used to characterize the solution behavior. The key finding is that while a variety of vesicle sizes ranging from 200 – 5000 nm in diameter can be formed based on initial processing conditions (including simple dissolution in buffer, extrusion/dialysis from organic solvents, and forming vesicles from polymer films deposited on surfaces) manipulation of the lysine block's secondary structure can be used to reversibly generate vesicles of well-defined size implying that the vesicles reported here are thermodynamic in nature. Moreover, the vesicles thus obtained can be reversibly formed after dilution and subsequent concentration of the solution, and even after precipitation/aggregation of the polypeptides by raising the pH to induce the coil-helix transition. The results of this work show that these previously unreported materials have unique properties of relevance to numerous applications including drug delivery, controlled released, encapsulation, and biomineralization/biomimetic syntheses of hard matter.

606c Release Behavior Of Paclitaxel From Self-Assembled Degradable Nanoparticles

Angel Mercado, Xuezhong He, and Esmaiel Jabbari. Chemical Engineering, University of South Carolina, Dept. of Chemical Eng., Rm. 2C11, Swearingen Engineering Center, Columbia, SC 29208

Introduction: Tumor vasculature has enhanced permeability regulated by factors such as VEGF and MMPs [2] with a pore cut-off size of 400 nm [3]. Due to the absence of an well-defined functioning lymphatic network, the clearance of macromolecules from the interstitial compartment is seriously compromised. This results in the accumulation of molecules with a molecular weight above 50 kDa in the interstitium and an outward convective interstitial flow. Bioresorbable nanoparticles are especially attractive as a delivery system because they can be degraded by hydrolysis or enzymatic pathways to low molecular weight compounds, thereby reducing the accumulation of nanoparticles in the interstitium. Our laboratory has developed novel bioresorbable macromers that self-assemble into nanoparticles in the aqueous medium. The macromer, hereafter abbreviated by PLGEOF, contains short lactide-co-glycolide (LG) and ethylene oxide (EO) blocks linked by fumaric acid units. After self-assembly, these nanoparticles can be crosslinked or grafted with cell-responsive biomolecules for targeted delivery of antitumor drugs. The objective of this study was to determine the size distribution, degradation profile, and release characteristics of the PLGEOF nanoparticles. Methods: A 10% w/w solution of the macromers in dimethyl sulfoxide (DMSO) was dialyzed against distilled deionized (DDI) water to obtain a stable colloidal suspension of nanoparticled. The morphology of the nanoparticles was examined using a JEOL 100 CXII transmission electron microscope at an accelerating voltage of 100 kV. The size distribution of nanoparticles was measured by dynamic light scattering with a NICOMP Submicron Particle Sizer (Autodilute Model 370). Degradation was measured as a function of time in vitro in primary culture media (CM; 5 ml per sample) without fetal bovine serum (FBS) at 37°C under mild agitation. Dextran fluorescein isothiocyanate (FITC-dextran) and paclitaxel were used as surrogate molecules for release studies. Results: The size of the nanoparticles depended strongly on the ratio of PLA to PLEOF in the mixture. The 100:0 90:10, 80:20, and 60:40 ratios, had number average size of 480, 195, and 50, and 14 nm and volume average size of 1630, 670, 185, and 23 nm, respectively. The nanoparticles, with the exception of 60:40 had bimodal size distribution. There was a significant decrease in the size of the particles from 480 to 195 and 50 nm as the PLAF:PLEOF ratio was varied from 100:0 to 90:10 and 80:20. FTIC-Dextran and paclitaxel released in a period of two weeks which matched the degradation profile of the spheres. CONCLUSION: Unsaturated terpolymer of lactide, ethylene oxide, and fumaric acid (PLEOF) with lactide to ethylene oxide ratios in the 90:10 to 70:30 spontaneously self-assembled into stable nanoparticles. The lowest particles size naoparticles was produced with lactide to ethylene oxide ratio of 60 to 40 (14 nm). These nanoparticles are potentially useful as a cell-responsive targeted delivery system for anitumor drugs.

606d Multi-Compartment Carriers For Enhanced Drug Delivery

Guohui Wu, Department of Chemical Engineering, University of California, Santa Barbara, Santa Barbara, CA 93106 and Joseph A. Zasadzinski, Department of Chemical Engineering, University of California, Santa Barbara, CA 93106.

Vesosome, a multi-compartment structure consisting of drug-loaded liposomes encapsulated within another bilayer, is a promising drug carrier with better retention and stability compared with unilamellar liposomes. The vesosome preparation takes advantage of the interdigitated phase of saturated lipids, which causes lipid bilayers to form flat, open sheets at low temperature, that close to form large unilamellar vesicles at higher temperatures. Adding poloxamer 188 (P188) to the Dipalmitoylphosphocholine (DPPC) interdigitated sheets prior to closure leads to ~ 0.6 μm diameter vesosomes. Applying osmotic pressure by diluting the vesosomes with buffer of increased salt concentration can further decrease the vesosome diameter by 10%. Freeze-fracture TEM and light scattering were used to evaluate the size distribution and polydispersity of the vesosomes formed. Nearly monodisperse hollow gold nanospheres with absorbance wavelength in the visible and infrared range have been synthesized and loaded inside vesosome to be externally triggered via near Infra-Red light for controlled drug release.

606e Controlling The Formation Of Multi-Compartment Vesicle Superstructures Using Membrane-Anchored DNA As Biomolecular Combination Locks

Paul A. Beales and T. Kyle Vanderlick. Department of Chemical Engineering, Princeton University, Engineering Quadrangle, Princeton, NJ 08544

Lipid vesicles are commonly used as model systems for biological cell membranes but they also have engineering applications as soft, technological containers. We study the formation of superstructures of vesicles with single-stranded DNA (ssDNA) anchored to their outer monolayers. Heteroaggregation is observed between vesicles decorated with ssDNA of a given sequence and vesicles decorated with the complimentary strand. We investigate how the growth behaviour of these vesicle assemblies is affected by several variables including vesicle concentration, vesicle charge and the average number of ssDNAs per vesicle. Three regimes of aggregation behaviour are observed: no aggregation, stable cluster sizes and continuous aggregation. The growth kinetics is determined by the collision rate of the vesicles in solution and the probability that a binding event will occur during a collision. The binding of short DNA oligonucleotides is thermally reversible: repeated assembly and disassembly of the vesicle aggregates can be induced by cycling in temperature. DNA binding is observed to be thermodynamically more stable when anchored to the vesicles than in bulk solution; we propose that this is due to the enhanced local concentration of DNA and a reduction in conformational entropy due to being anchored to the membrane surface. We also demonstrate that this assembly technique is applicable to vesicles over a wide range of length-scales from vesicle populations with an average diameter of approximately 80nm to giant vesicles with diameters greater than 5μm.

Packaging of a Polymer by Viral Nanocontainer

Yufang Hu[1], Roya Zandi[2], Adriana Anavitarte[3], Charles M. Knobler[3], and William M. Gelbart[3]
1, Department of Molecular and Medical Pharmacology and Department of Urology, David Geffen School of Medicine at UCLA. 2, Department of Physics, University of California, Riverside. 3, Department of Chemistry and Biochemistry, University of California, Los Angeles.

Introduction Recently viral capsids have emerged as a new class of biomaterials for the creation of novel biomimetic materials in applications ranging from biosensors to displays to drug delivery (1-4). Many of these applications take advantage of the inherent symmetry and uniform size of viral capsids and use them as building blocks of more sophisticated hierarchical structures. A key issue that arises from these applications is the determination and control of viral capsid size and shape: Is there a correlation between the genome size and the capsid size of these spherical RNA viruses? This paper attempts to address some of these questions. We focus our attention on a spherical, single-stranded RNA virus called the Cowpea Chlorotic Mottle Virus (CCMV). The viral particle formation is driven by interactions between the anionic viral RNA and the cationic capsid protein residues (5, 6). The pioneering work of Bancroft and coworkers demonstrated the ability of CCMV capsid protein (CP) to encapsidate a wide range of non-viral anionic materials (7, 8). In recent years, several groups have focused on using CCMV and other members of the *Bromoviridae* virus family for biomaterials applications (2, 9-13). We report results from *in vitro* self-assembly study of the CCMV CP and a linear anionic polyelectrolyte, polystyrene sulfonate (PSS). We chose to replace viral RNA with PSS in our study, whose solution size and shape properties are well documented. PSS of five molecular weights ranging from 400 kDa to 3.4 MDa were packaged in independent self-assembly reactions by CCMV CP. The morphology and number density of the resulting VLPs from each reaction were characterized by transmission electron microscopy (TEM). The goal of the experiments was to probe whether there is a particular size of PSS for which the packaging was optimum, and if so, to elucidate the physical principles underlying the correlation between the PSS molecular weight (M_w), capsid size, and packaging efficiency.

Experimental methods All PSS samples were purchased from Scientific Polymer Products, Inc. (Ontario, NY) and were used without further purification. Table 1 lists the molecular weights, polydispersity indexes, and measured hydrodynamic radii for the PSS samples used in our experiments. Five PSS samples with M_w ranging from 400 kDa to 3.4 MDa were used.

Table 1. Properties of PSS used in packaging experiments

Reaction	M_w (Da)	Polydispersity[1] index	Hydrodynamic diameter[2] (nm)
1	408,400	1.24	36.5
2	691,300	1.16	50.0
3	1,010,000	1.18	65.1
4	2,040,000	1.07	73.2
5	3,441,000	1.01	85.6

Each experiment consisted of five independent self-assembly reactions carried out simultaneously, one for each M_w of PSS listed in Table 1. The CP concentration in each reaction was held constant at 0.15 μg/μL, and the molar ratio of CP to PSS in all reactions was 360:1. The reaction mixtures were dialyzed against the assembly buffer for 24 hr (50 mM sodium chloride, 10 mM potassium chloride, 5 mM magnesium chloride, 1 mM DTT, 50 mM Tris-hydrochloride, pH 7.2).

Results Figure 1a shows the representative TEM image from the five packaging reactions listed in Table 1. For comparison, images of the empty CCMV capsid and wild-type CCMV (wt-CCMV) are also shown. VLPs with spherical morphology formed readily in Reactions 1-5. Stain penetration and accumulation inside the capsid is clearly visible for the empty CCMV capsids (Figure 1b), as characterized by the presence of prominent dark centers in all capsids. Similarly, the wt-CCMV capsids also display a certain degree of stain penetration (Figure 1c), albeit less than that in the empty CCMV capsids. The VLP capsids from assembly reaction in Figure 1a however, mostly lacked any significant stain penetration. The morphological differences between empty capsids, wt CCMV, and PSS VLPs as shown in Figure 1 reflected qualitatively the volume occupied by the packaged cargo inside these capsids. Further, we believe that most capsids formed in our assembly reactions contained encapsidated PSS, which excluded the stain. "Filled" capsids such as those shown in Figure 1a do not form with CP alone near neutral pH, nor from PSS alone without any CP: the formation of empty CCMV capsids at neutral pH is hindered by electrostatic repulsion from the highly positively charged CP N-termini. Capsid formation at neutral pH is facilitated by the presence of polyanions such as viral RNA or PSS, which stabilizes the Coulomb attractions between the polyanions and the protein N-termini (5, 6).

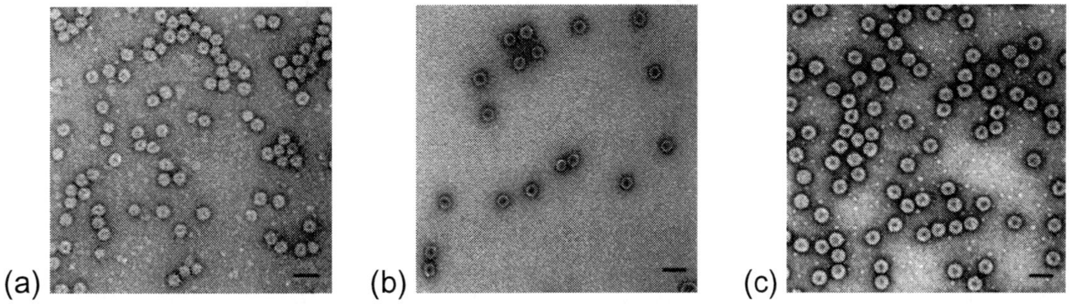

Figure 1. TEM images of capsids. (a) VLPs formed with 2 MDa PSS. (b) Empty CCMV capsid formed by dialysis of CP in buffer with high salt and low pH. (c) wt-CCMV capsids in virus suspension buffer. Scale bars are 50 nm.

The normalized VLP capsid size distributions and their Gaussian fits for Reactions 1-5 are shown in Figure 2. The size distribution for each individual reaction is uni-modal. The combined normalized size distribution was obtained by summing up ~600 particles from Reactions 1-5 and performing Gaussian fitting on the total particle population. The Gaussian fit of the total size distribution are bi-modal with peaks at 21.5 and 27.3 nm, respectively. Nearly all capsid sizes measured fall into the range of 16-34 nm. A summary of the Gaussian peak fitting result is given in Table 2.

Table 2. Summary of the Gaussian fit of the VLP size distributions from Reactions 1-5.

Reaction	PSS Mw (Da)	Peak	
		Position (nm)	Width (nm)
1	408,400	21.1	1.6
2	691,300	22.3	1.6
3	1,010,000	22.0	2.2
4	2,040,000	27.1	1.8
5	3,441,000	28.3	2.0

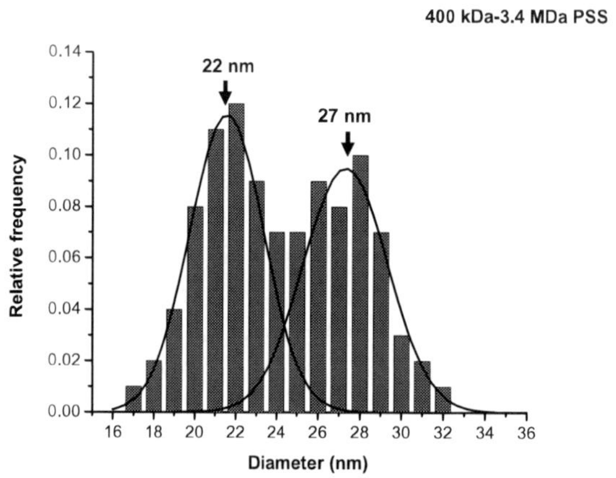

Figure 2. A combined normalized capsid size histogram of VLPs from all five reactions. A fit of the histogram to two Gaussians indicates that the capsid sizes converge to two values: 22 nm and 27 nm. These values agree well with the sizes of T=2 and T=3 capsids formed by CCMV capsid proteins.

Figure 2 and Table 2 above clearly demonstrate that the VLPs formed in all five reactions fall into two size groups with their average capsid diameters around 22 and 27 nm, respectively. The fact that the sizes of hundreds of VLP capsids from five independent packaging reactions converge to these two discrete values strongly implies that there exists an intrinsic hierarchical order in the capsid structure, which we believe is due to the inherent icosahedral symmetry of these viral capsids. Further, we believe that these two populations of capsids correspond to T=2 and T=3 icosahedral structures based on a large body of literature on VLPs formed by CCMV CP. It has been established that the CCMV CP is capable of forming a number of icosahedral structures with T=1, 2, 3, 4, and 7 (7, 8, 14, 15). The type of structure formed is typically functions of both the solution condition and the properties of the anionic template. Among these, capsids with diameters of 22 and 28 nm have been observed via negative staining TEM and have been associated with "T=2" and T=3 icosahedral symmetries, respectively (2, 15, 16). Strictly speaking, the "T=2" capsids do not follow the *quasi-equivalent* rule as proposed by Casper and Klug (17) because the capsid is not composed of CP pentemers and hexamers but of "pentamers-of-dimamers" (14, 15). The structure of "T=2" capsids formed by CCMV CP has recently been resolved by Cryo-TEM (15). Unequivocal determination of the structure of the VLP capsids formed in our reactions requires techniques such as cryo-TEM and X-ray crystallography, which are beyond the scope of this study.

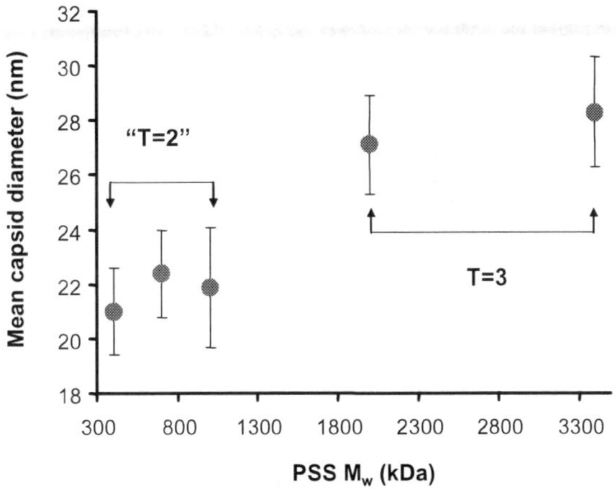

Figure 3. The mean capsid size as a function of packaged PSS M_w. A jump in the capsid size was observed when the M_w increased from 1 MDa to 2 MDa.

Figure 3 shows the mean capsid diameter as a function of PSS molecular weight. A jump in the capsid symmetry from "T=2" to T=3 occurred between PSS Mw 1M and 2M. The switch of the capsid size into larger T number strongly suggests that for each type of capsids ("T=2" or T=3) there exists a maximum packing capacity: As the packaged cargo size increases, the size of the capsid increases correspondingly. Because the icosahedral symmetry has been inherently built into the CP, the capsid size transition as observed in our experiments is discrete, rather than continuous.

Discussions If the preference of the CPs to aggregate with a particular radius of curvature were sufficiently strong, there would only be a single VLP size regardless of the PSS M_w and the overall and relative concentrations of CP and PSS. The switch to larger VLP size upon increase in PSS M_w as found in our experiments can be understood by a simple model incorporating the physics of capsid self-assembly and polymer confinement. This model reflects two basic criteria for the formation of a stable capsid: First, for a given PSS size, the capsid should be sufficiently large so the self-repulsion energy and entropy of the charged PSS due to confinement are not too energetically costly. Second, the capsid should not be too large such that the PSS is not able to neutralize the positively charged amino acid residues on the inner capsid surface without strong absorption onto that surface and attendant loss of configurational entropy.

The free energy of the VLP in our model includes: (1) the self-energy of the CP capsomer, i.e, protein pentamers and hexamers, E_1; (2) the attraction between PSS and the inside of the shell, E_2; (3) the self-energy of the PSS chain, E_3; (4) the configurational entropy of the PSS chain, E_4:

$$E_{confined\,chain} = E_1 + E_2 + E_3 + E_4 . \qquad (2)$$

Because the screening length under our assembly buffer condition is comparable to the PSS monomer size, both are on the order of 1nm, E_2 in Eq (2) can be described by a "contact" energy per monomer ($-\varepsilon$). E_3 can be treated via the usual monomer excluded-volume interaction and is related to the second virial coefficient. The fact that the PSS chain is flexible with its persistence length being comparable to its monomer size allows the chain entropy contribution E_4 to be described by the familiar gradient-squared term with coefficient varying inversely with monomer size and volume fraction (cf. Ji and Hone). Finally, the self-energy of the CP aggregate can be treated by a simple Hamiltonian developed in our earlier work to explain the origin of icosahedral symmetry in viral capsids (18).

Figure 4 illustrates schematically (dotted curve) the self energy E_1 per capsomer as a function of the total number of capsomers in a capsid, N, for a range of N values bracketing two neighboring T structures, e.g., T=2 and T=3. This is the type of energy dependence that is calculated in a model treating CP hexamers and pentamers as Lennard Jones disks whose centers are restricted to lie on the surface of a sphere of radius R_c. Monte Carlo simulation provides the equilibrium spatial distributions of capsomers for each N and R_c, with subsequent minimization with respect to R_c giving the energy per capsomer for each N(18). As N approaches infinity, E_1 per capsomer levels off asymptotically to a value corresponding to that of a planar hexagonal array of CPs. Note that the two minima shown in the figure are only separated by a small energy difference, $\Delta\varepsilon$. Because this difference is often not large compared to the thermal energy kT, one expects to observe both structures with their relative populations given by the corresponding Boltzmann factor $[\exp(-\Delta\varepsilon/kT)]$.

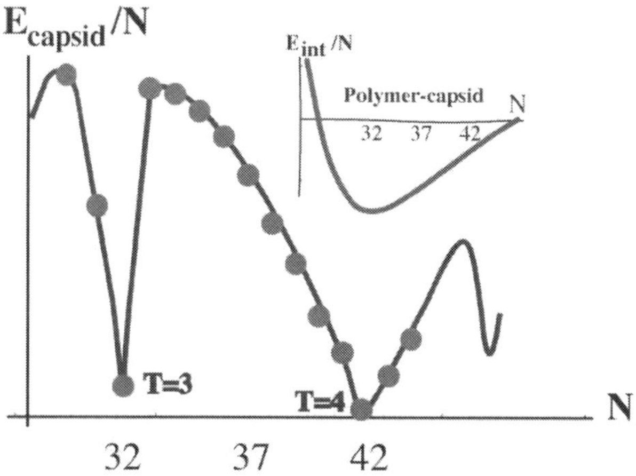

Figure 4. Illustration of the VLP size selection mechanism by PSS M_w. The dotted curve shows the self-energy per capsomer in an empty capsid as a function of the number of capsomers per capsid. The energy minima correspond to structures with icosahedral symmetry (N=32, 42), whose formation is energetically more favorable. The smooth curve in the insert shows the interaction energy per capsomer of a given M_w PSS with the capsid as a function of capsid size. For a given PSS molecule, the interaction energy goes through a minimum as the capsid size varies. Thus, there exists an optimal capsid size for the PSS and the capsid size is determined by superimposition of the two energy curves.

Note that the above calculations take into account only the protein-protein interactions. In the absence of PSS or RNA, empty CCMV capsids do not arise under physiological conditions, since the energy minima corresponding to T=2 and T=3 empty capsids as shown in Figure 4 are not sufficiently low compared to the chemical potential of *dis*aggregated CP subunits in solution. To drive the capsomer aggregation and the spontaneous assembly of capsids it is necessary to reduce the pH and raise the ionic strength or to add an anionic polymer (such as PSS or RNA) or an anionic colloidal particle (such as functionalized gold nanoparticles and quantum dots) (7, 13). The energies of confining the PSS chain inside the viral capsid are given by terms E_2, E_3 and E_4 in Eq (2). The treatment of these terms is analogous to the case of a semi-dilute polymer solution confided inside a spherical cavity to which the polymer is attached, as give by Ji and Hone (19). The polymer is flexible and of a fixed length. For particular values of the PSS monomer-capsid contact energy $-\varepsilon$, monomer-monomer energy, and chain length, a minimum in interaction energy per unit capsid surface area exists as a function of the capsid radius. This is shown schematically by the smooth curve in Fig. 4, in a plot of PSS-capsid interaction energy per unit area of the capsid interior surface as a function of the total capsomer number N, which is proportional to the capsid radius. The position and depth of the minimum depend quantitatively on the values of the monomer excluded volume, monomer-capsid contact energy, and PSS length, but the existence of a preferred capsid size as found in our experiments is a qualitative result of the balance between the PSS chain self-repulsion and its attraction to the interior surface of the confining capsid.

Combining the two (smooth and dotted) curves in Fig. 4 by adding the capsid self-energy to the contributions from the confined PSS chain results in the solid curve. Note that, for the case shown here the chain contribution "selects" the smaller capsid size. For a longer PSS chain, the minimum in the smooth curve would shift towards larger N, thereby favoring the larger capsid. This is consistent with our finding that VLP size is shifted to larger T numbers by increasing the PSS molecular weight.

References

1. Garcea, R. L., L. Gissmann. 2004. Virus-Like Particles as Vaccines and Vessels for the Delivery of Small Molecules. Current Opinion in Biotechnology 15: 513.
2. Sun, J., C. DuFort, M. C. Daniel, A. Murali, C. Chen, et al. 2007. Core-Controlled Polymorphism in Virus-Like Particles. Proceedings of the National Academy of Sciences of the United States of America 104: 1354.
3. Lewis, J. D., G. Destito, A. Zijlstra, M. J. Gonzalez, J. P. Quigley, et al. 2006. Viral Nanoparticles as Tools for Intravital Vascular Imaging. Nature Medicine 12: 354.
4. Stray, S. J., J. M. Johnson, B. G. Kopek, A. Zlotnick. 2006. An in Vitro Fluorescence Screen to Identify Antivirals That Disrupt Hepatitis B Virus Capsid Assembly. Nature Biotechnology 24: 358.
5. Speir, J. A., S. Munshi, G. J. Wang, T. S. Baker, J. E. Johnson. 1995. Structures of the Native and Swollen Forms of Cowpea Chlorotic Mottle Virus Determined by X-Ray Crystallography and Cryoelectron Microscopy. Structure 3: 63.
6. Fox, J. M., G. J. Wang, J. A. Speir, N. H. Olson, J. E. Johnson, et al. 1998. Comparison of the Native Ccmv Virion with in Vitro Assembled Ccmw Virions by Cryoelectron Microscopy and Image Reconstruction. Virology 244: 212.
7. Bancroft, J. B., Hiebert, E., Bracker, C. E. 1969. The Effects of Various Polyanions on Shell Formation of Some Spherical Viruses. Virology 39: 924.
8. Verduin, B. J. M., J. B. Bancroft. 1969. Infectivity of Tobacco Mosaic Virus Rna in Coat Proteins from Spherical Viruses. Virology 37: 501.
9. Douglas, T., M. Young. 1998. Host-Guest Encapsulation of Materials by Assembled Virus Protein Cages. Nature 393: 152.
10. Douglas, T., M. Young. 1999. Virus Particles as Templates for Materials Synthesis. Advanced Materials 11: 679.

11. Gillitzer, E., D. Willits, M. Young, T. Douglas. 2002. Chemical Modification of a Viral Cage for Multivalent Presentation. Chemical Communications: 2390.
12. Dixit, S. K., N. L. Goicochea, M. C. Daniel, A. Murali, L. Bronstein, et al. 2006. Quantum Dot Encapsulation in Viral Capsids. Nano Letters 6: 1993.
13. Chen, C., M. C. Daniel, Z. T. Quinkert, M. De, B. Stein, et al. 2006. Nanoparticle-Templated Assembly of Viral Protein Cages. Nano Letters 6: 611.
14. Zlotnick, A., R. Aldrich, J. M. Johnson, P. Ceres, M. J. Young. 2000. Mechanism of Capsid Assembly for an Icosahedral Plant Virus. Virology 277: 450.
15. Tang, J. H., J. M. Johnson, K. A. Dryden, M. J. Young, A. Zlotnick, J. E. Johnson. 2006. The Role of Subunit Hinges and Molecular "Switches" in the Control of Viral Capsid Polymorphism. Journal of Structural Biology 154: 59.
16. Krol, M. A., N. H. Olson, J. Tate, J. E. Johnson, T. S. Baker, P. Ahlquist. 1999. Rna-Controlled Polymorphism in the in Vivo Assembly of 180-Subunit and 120-Subunit Virions from a Single Capsid Protein. Proceedings of the National Academy of Sciences of the United States of America 96: 13650.
17. Caspar, D. L. D., Klug, A. 1962. Physical Principles in the Construction of Regular Viruses. Cold Spring Harb. Symp. Quant. Biol. 27: 1.
18. Zandi, R., D. Reguera, R. F. Bruinsma, W. M. Gelbart, J. Rudnick. 2004. Origin of Icosahedral Symmetry in Viruses. Proceedings of the National Academy of Sciences of the United States of America 101: 15556.
19. Ji, H., D. Hone. 1988. Polymer Adsorption on Rough Surfaces .2. Good Solvent Conditions. Macromolecules 21: 2600.

606g Molecular Recognition of Glycolipid Biosurfactants toward Various Immunoglobulins

Seya Ito[1], Tomohiro Imura[2], Fukuoka Tokuma[2], Tomotake Morita[2], Hideki Sakai[1], Masahiko Abe[1], and Dai Kitamoto[2]. (1) Department of Pure and Applied Chemistry, Faculty of Science and Technology, Tokyo University of Science, 2641 Yamazaki, Noda, Chiba 278-8510, Japan, (2) Research Institute for Innovation in Sustainable Chemistry, National Institute of Advanced Industrial Science and Technology (AIST), Tsukuba Central 5-2, Higashi 1-1-1, Tsukuba, Ibaraki 305-8565, Japan

Biosurfactants are amphiphilic compounds produced by microorganisms, and have been receiving increasing attention due to their unique properties compared to petroleum-based surfactants. Mannosylerythritol lipids (MELs), which are efficiently produced from vegetable oils by yeasts, are one of the most promising biosurfactants known [1, 2]. Recently, we have demonstrated that MEL-A, the major component of MELs, exhibits not only excellent surface-active and self-assembling properties [3] but also high binding affinity towards glycoproteins such as immunoglobulins and lectins [4]. However, available information on the binding manner is still limited. Considering the development of MEL-A into a new ligand for immunoglobulins, it is of great interest to characterize the molecular recognition further in detail. In this study, the binding between the self-assembled monolayer prepared from MEL-A and different immunoglobulins (IgG, IgM, and IgA) was investigated by a surface plasmon resonance (SPR) technique and an *in-situ* atomic force microscopy (AFM). We found that the glycolipid monolayers exhibit a high binding affinity towards all the immunoglobulins tested. The calculated binding constants of MEL-A with IgG, IgM and IgA were 9.4×10^6 M^{-1}, 5.4×10^6 M^{-1}, 7.2×10^6 M^{-1}, respectively. Moreover, we succeeded in directly observing a large amount of immunoglobulin molecules bound to the monolayer with a markedly high density. Thus, MEL-A is very likely to have a great potential to serve as a novel affinity ligand on sensing and/or separation of immunoglobulins and related glycoproteins.

[1] Kitamoto *et al.*, J. Biosci. Bioeng., **94**, 187 (2002). [2] Morita *et al.*, Appl. Microbiol. Biotechnol., **73**, 305 (2006). [3] Imura *et al.*, Langmuir, **23**, 1659 (2007). [4] Konishi *et al.*, Biotechnol. Lett., **29**, 865 (2007).

606i Grazing Incidence Small Angle X-Ray Scattering Characterization Of 2D Self-Assembled Bacteriophage Arrays Deposited Via A Convective Transport Process

Carlee E. Ashley[1], Eric C. Carnes[1], Landon T. White[1], Zhen Yuan[1], Darren R. Dunphy[2], Dimiter Petsev[1], Plamen Atanassov[1], David Peabody[3], and C. Jeffrey Brinker[2]. (1) Department of Chemical Engineering, University of New Mexico, Advanced Materials Laboratory, 1001 University Blvd. SE, Suite 100, Albuquerque, NM 87106, (2) Sandia National Laboratories, 1001 University Blvd. SE, Suite 100, Albuquerque, NM 87106, (3) Department of Molecular Genetics and Microbiology, University of New Mexico, Advanced Materials Laboratory, 1001 University Blvd. SE, Suite 100, Albuquerque, NM 87106

Bacteriophages are viruses that infect and replicate in bacteria and possess several characteristics that make them ideally suited for use in studying nanoparticle self-assembly into hierarchical structures. Phage populations naturally possess a high degree of monodispersity and occur in a variety of geometries similar to inorganic nanoparticles (e.g. spheres, anisotropic rods, etc.). Additionally, bacteriophages have highly regular, symmetrical geometries and dimensions from tens to hundreds of nanometers, enabling their self-assembly into ordered arrays via evaporation-driven convective transport. The self-assembly of bacteriophages with tailorable geometry, size, and surface properties into highly-ordered 2D and 3D arrays can be used to create new nanoparticle architectures with unique properties. We have used two 28-nm icosahedral coliphages, MS2 and Qβ, in conjunction with a novel deposition technique, convective assembly, to study bacteriophage self-assembly into mono- and multilayers. Convective assembly enables rapid deposition of colloidal particles onto a hydrophilic surface by trapping microliter-sized droplets between two fixed-angle plates and dragging the ensuing meniscus across a substrate with constant velocity. Deposition speed, relative humidity, and particle volume fraction can be varied to control film structure and thickness. Convective assembly is well-suited for the deposition of biological species, as it does not require the use of volatile co-solvents and minimizes material consumption. We have employed Grazing Incidence Small Angle X-ray Scattering (GISAXS) at a synchrotron source to study the time-resolved self-assembly of MS2 and Qβ at various deposition velocities, in solvents with different ionic strengths, and onto a variety of substrates. GISAXS is a preferable method to assess the structure of bacteriophage and nanoparticle arrays as it provides detailed information about the long-range order of the bulk film, whereas electron microscopy techniques only provide structural information about isolated, submicrometer-sized areas. We have found that particle-particle and particle-substrate interactions greatly influence the structure, thickness, and long-range order of bacteriophage arrays, as well as the dynamics of film formation. MS2, which has localized regions of positive and negative charges on the surface of its capsid, forms randomly-oriented mono/multilayers, while Qβ, which possesses a more disperse surface charge density, forms highly-ordered hexagonal-close packed (HCP) monolayers. The ultimate structure and long-rage order of bacteriophage arrays are highly influenced by the charge density of the hydrodynamic particle, which can be varied by changing the solvent ionic strength or by substitution of surface amino acids in the phage capsid. We have assessed the affect of charge density on MS2 and Qβ arrays by varying the ionic strength of the buffer; zeta potential and atomic force microscopy measurements have been used to supplement GISAXS data. We have, furthermore, characterized the influence of particle-substrate interactions on long-range order using hydrophobic, anionic, cationic, and non-ionic, polar surfaces and by employing 10% (v/v) glycerol as the solvent, which allows the particles to freely diffuse on the substrate. To assess the role of particle size on the convective assembly of colloids, we have used the globular protein ferritin, which is half the size of MS2 and Qβ, and have found that it assembles into disordered mono/multilayered structures. Furthermore, we have assembled phages with different geometries, such as tobacco mosaic virus, which is anisotropic, and T4 bacteriophage, which has an icosahedral head and an anisotropic tail, onto lithographically-patterned substrates with submicrometer-sized features to test whether phages with anisotropic geometries preferentially organize perpendicular to or parallel to the patterned nanochannels.

617a Magnetite Nanoparticles for Magnetic Drug Targeting

Jan O. Mangual, Misael O. Aviles, Armin D. Ebner, and James A. Ritter. Department of Chemical Engineering, University of South Carolina, Columbia, SC 29208

Ritter and co-workers have been exploiting nanostructured magnetic materials as a scaffold for implant assisted - magnetic drug targeting (IA-MDT). In MDT, an external magnetic field is used to attract and retain magnetic drug carrier particles (MDCPs) at a specific site in the body. However, the use of an external field alone is limited by the small magnetic forces exerted on the MDCPs, which leads to poor retention. IA-MDT represents a possible solution to this dilemma that Ritter and co-workers have been exploring. One IA-MDT approach they have been testing uses magnetic nanoparticles, such as magnetite, to enhance the capture of the MDCPs at the target site. These magnetite nanoparticles, which are far more magnetic than the MDCPs, are easily collected at the target site when placed in the magnetic field. So they are collected first, followed by the MDCPs. This approach works because the magnetic nanoparticles become magnetized in the magnetic field and thus impart a force on the MDCPs that is larger than that generated by the magnetic field alone. In effect, the magnetic nanoparticles act as a scaffold for the MDCPs to be attracted to, with both being collected and retained magnetically at the target site. This IA-MDT system was studied in vitro using a porous polyethylene cylinder to simulate capillary tissue. The magnetite nanoparticles were first captured within the porous polymer using an external magnet. They ranged in diameter from 10–100 nm. The MDCPs were then captured using the magnetite nanoparticles as a magnetic scaffold or implant, as explained above. The MDCPs were 0.87 ìm in diameter and embedded with 25 wt% magnetite. This presentation will disclose the effects of several variables on the performance of this unique IA-MDT approach in terms of the MDCP capture efficiency (CE). The fluid velocity, the distance to the magnet, the applied magnetic field, the magnetite nanoparticle size and concentration, and the MDCP concentration were studied. The results showed a significant increase in the capture of the MDCPs surrogates when the magnetite nanoparticles were present. The results also revealed considerable insight to the design of such an IA-MDT system.

617b Functionalization Of Hydrophilic Magnetite Nanoparticles With Intelligent Hydrogels Via Atrp

Reynolds A. Frimpong and J. Zach Hilt. Chemical and Materials Engineering, University of Kentucky, 177 F. Paul Anderson Tower, Lexington, KY 40506

The intrinsic magnetic properties of iron oxide (Fe3O4) have been exploited in various applications. Particularly for biomedical applications, their effectiveness is enhanced with suitable coatings on their surface. Most of the current syntheses of Fe3O4 involve the use of organic surfactants such as oleic acid and oleylamine which provide stable dispersions only in organic solvents. Stable hydrophilic nanoparticles which are critical for certain applications are only obtained after subsequent post processing reactions to exchange the hydrophobic surfactant coating. Here, we describe a one-step co-precipitation synthetic approach to obtain citric acid coated Fe3O4 nanoparticles and the functionalization of these nanoparticles with intelligent hydrogels (e.g. thermo-responsive N-isopropylacrylamide) via surface-initiated atom transfer radical polymerization (ATRP). Hydroxyl groups attached to the Fe3O4 nanoparticles after co-precipitation result in magnetic nanoparticles which are stable in aqueous systems, and further provide initiating sites for the ATRP reaction. The nanoparticles are characterized for their size using both dynamic light scattering and transmission electron microscopy. The coating is also confirmed by fourier transform infrared spectroscopy and thermo-gravimetric analysis.

617d Functionalization of Superparamagentic Iron Oxide Nanoparticles (Spion) for Fluorescent and MRI Imaging and Its Application for Cell Labeling

Jae-Ho Lee and Joseph Frank. Experimental Neuroimaging Section, National Institutes of Health, Clinical Center NIH Building 10 Room B1N256, Bethesda, MD 20892-1074

Superparamagentic iron oxide nanoparticles (SPION) are a class of magnetic resonance imaging (MRI) contrast agent. Dual or multimodal imaging based on SPION requires more for non-destructive, fast, and accurate imaging analysis of tumor margin, and temporal and spatial migration of stem cells. Several groups have reported functionalization of dextran-coated SPION but not with detailed and simple approach. In this study, with a simple oxidation approach we have introduced functional groups into dextran coated SPION and reacted with several different fluorescent dyes for fluorescent and MRI imaging. The fluorescent SPION were stable and successfully used for cell labeling. Briefly cell labeling was conducted using fluorescent dextran-coated SPION and protamine sulfate, conventionally used to reverse heparin anticoagulation. After labeling of hela cell and human mesenchymal stem cells (MSCs) with fluorescent dextran-coated SPION such as fluorescent ferumoxides-protamine sulfate complexes (FLUO FE-Pro), quantitative cellular iron incorporation was determined. FLUO FE-Pro cell labeling was almost the same as the previous FE-Pro labeling. Fe-Pro labeled cell showed stronger fluorescent image, which confirmed successful fluorescent dye conjugation onto dextran-cated SPION.

617e Application of Magnetic Nanoparticle Embedded Nanofibers in Bio-Fuel Production

S.M. Habibur Rahman, Environmentally Sustainable Process Technology, Chalmers university of Technology, SE-412 96, Göteborg, Sweden, Muhammad Moniruzzaman, Department of Applied Chemistry, Kyushu University, 744 Moto-oka, Fukuoka, Japan, and **Mohammad Munim Hussain**, MGP Ingredients, Peoria, IL 61615.

The emergences of nanotechnology opened up so many research areas and electrospun nanocompositefiber is a potential candidate that could be used as an efficient cellular scaffold. It is possible to fabricate functionally active nanofibers based on the end-use applications. In fermentation, magnetic nanoparticles play role to increase the rate of oxygen transport. The goal of this research is to fabricate biodegradable nanofibers embedded with magnetic nanoparticles. Nanofibers in the range of 300-500 nm embedded with magnetic nanoparticles have been fabricated successfully. These nanofibers have also been used in the fermentation process. The growth of yeast on the nanofibers has been examined and oxygen transport rate has been calculated. The results are then compared with a fermentation process without having nanofibers and magnetic nanoparticles. Significant improvements have been noticed which will be demonstrated in the final presentation.

617f Development of a Digital Microfluidic Lab-on-a-Chip for Automated Immunoassay with Magnetically Responsive Beads

637a Bio-Derived Antimicrobial Materials and Coatings

D. Matthew Eby[1], Heather R. Luckarift[2], and **Glenn R. Johnson**[2]. (1) Airbase Technologies Division, Air Force Research Laboratory, 139 Barnes Drive, Suite 2, Tyndall AFB, FL 32403, (2) AFRL/MLQL, Air Force Research Laboratory, 139 Barnes Drive, Suite # 2, Tyndall Air Force Base, FL 32403

Numerous biochemical mechanisms protect organisms against microbial pathogens. Materials that incorporate active antimicrobial biomolecules could provide a significant advance for self-sanitizing surfaces and decontamination processes. In the present work we have found that the bacteriolytic hydrolase, lysozyme, and the antimicrobial peptide, KSL, will catalyze in vitro mineralization of inorganic oxides from soluble salts. The reaction provides nanoporous particles, 50 to 500 nm in diameter; their size and morphology are influenced by the protein template and precursor solution used in the reaction. The biological activity of the template molecule can be retained when associated with the inorganic oxide matrices. Lysozyme shows remarkable retention of native bactericidal activity and protein structure when associated with silica. Similarly, KSL will promote mineralization reactions in vitro and demonstrates antimicrobial activity in the solid-state composite. In another approach, lysozyme catalyzed reduction of silver anions to form composites of silver nanoparticles and active lysozyme. The composite material could be transferred to stainless steel surfaces using electrodeposition to create an active antimicrobial coating. The silver composite material takes advantage of the antimicrobial effect of the silver metal, as well as the biochemical activity of the template protein.

637b Biofabrication With Genetically Modified Viral Nanotemplates

Hyunmin Yi[1], Gary W. Rubloff[2], Gregory F. Payne[3], William E. Bentley[4], and James N. Culver[3]. (1) Department of Chemical and Biological Engineering, Tufts University, 4 Colby St., Science & Technology Center Rm 147, Medford, MA 02155, (2) Institute for Systems Research, University of Maryland, 2145 A.V. Williams Bldg., College Park, MD 20742, (3) University of Maryland Biotechnology Institute, 5115 Plant Sciences Building, College Park, MD 20742, (4) University of Maryland, College Park, 6143 Plant Science Bldg, College Park, MD 20742

Biological macro/supramolecules such as DNA, proteins and viruses hold great promise as templates for nanodevice fabrication due to the inherent advantages over inorganic materials such as selectivity, mild processing conditions and precise self-assembly mechanisms nature has provided. Despite recent advances in discovering novel properties of various nanoscale entities, reliable fabrication of nanodevices remains challenging due to the lack of facile assembly techniques that allow high spatial and orientational control. To address this we have developed biochemically driven patterned assembly of potent viral nanotemplates onto inorganic substrates. Specifically, Tobacco Mosaic Viruses serve as high capacity nanotemplates via a genetically inserted cysteine on each coat protein, which provides covalent coupling sites. These functionalized nanotemplates were partially stripped and assembled onto capture DNA surfaces through hybridization with its own genomic RNA. Fluorescence and SEM images show high spatial, orientational and sequence specificity as well as stability of the patterned TMV templates . Further, simultaneous assembly of multiple TMVs onto printed DNA microarray platforms at readily addressable locations via linker DNA hybridization based programming is demonstrated. Our recent progress on hierarchical assembly of these potent nanoentities will also be presented.

637c Bioseparations Using A Nanoisland Array Made Of Self-Assembled Metal Oxides

Lawrence Burr Zimmerman, Chemical and Biomolecular Engineering, The Ohio State University, 125 Koffolt Labs, 140 W. 19th Ave., Columbus, OH 43210, L. James Lee, The Department of Chemical and Biomolecular Engineering, The Ohio State University, 125 Koffolt Labs, 140 W. 19th Ave., Columbus, OH 43210, Frantisek Svec, Department of Chemistry, University of California at Berkeley, 729 Latimer Hall, Berkeley, CA 94720-1460, and Michael. D. Rauscher, Materials Science and Engineering, The Ohio State University, 477 Watts Hall, 2041 College Rd., Columbus, OH 43210.

Our novel metal-oxide based nanostructures present a tunable, low-cost, robust platform for nanofluidic devices. We created an orthogonal network of nanochannels (50-200nm) and drove DNA through them via electrophoresis and capillary action. These dimensions distributed uniformly across large areas (25 mm2 currently, larger expected) are optimal for examining DNA, proteins, or other nanoparticles. We imaged individual molecules of DNA within the channels and demonstrate separation of DNA by size. We also characterized the surface of the nanoislands, made of a mixture of GDC (gadolinium doped ceria) and YSZ (yttria-stabilized zirconia), and developed means for modifying the naturally hydrophilic surface via UV-initiated polymer grafting. The nanoislands can be selectively treated based on surface chemistry and photolithography. Finally, we demonstrate pattern transfer to polymers to enable inexpensive polymeric nanofluidic devices. The nanoislands and substrate are much stronger than silicon wafers, making a superior mold material.

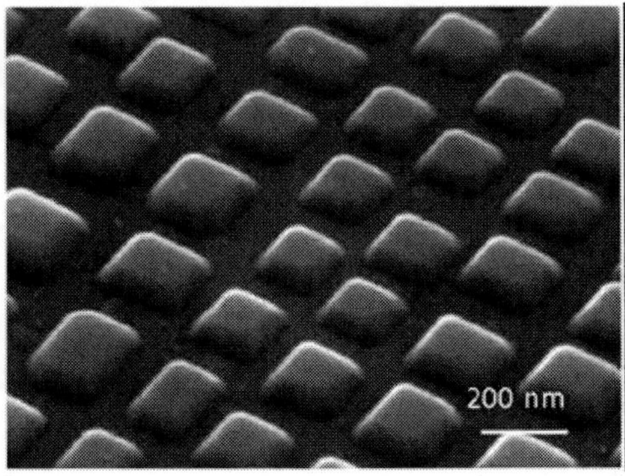

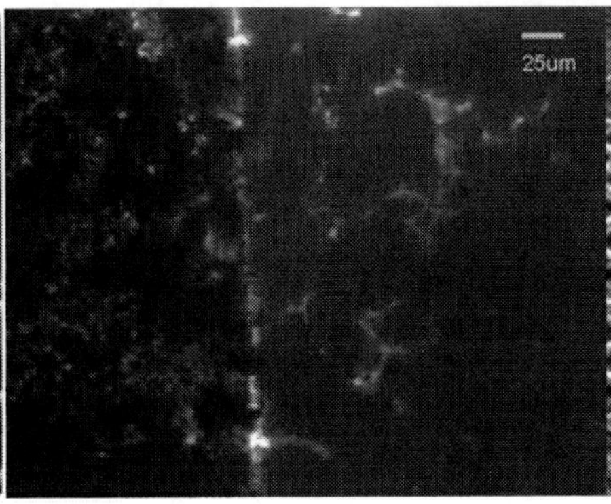

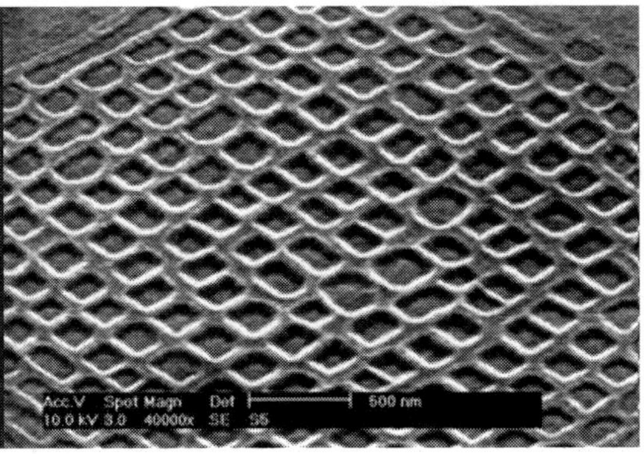

637d Polymer Systems Tuned To Human Blood Outgrowth Endothelial Cell Adhesion

Daniel Heath[1], Anka N. Veleva[2], Cam Patterson[3], John J. Lannutti[4], and Stuart L. Cooper[1]. (1) Department of Chemical and Biomolecular Engineering, The Ohio State University, Columbus, OH 43210, (2) Department of Chemical and Biomolecular Engineering, North Carolina State University, 911 Partners Way, Box 7905, Raleigh, NC 27695, (3) Department of Medicine, The University of North Carolina at Chapel Hill, 8200 Medical Biomolecular Research Bldg., 103 Mason Farm Road, Chapel Hill, NC 27599, (4) Department of Material Science and Engineering, Ohio State University, Columbus, OH 43210

Statement of Purpose: The function of many blood contacting biomedical devices is compromised by thrombus development at the biomaterial/blood interface. No surface except a healthy endothelium is fully blood compatible. However, a confluent and functioning endothelial cell layer on a biomaterial surface has not yet been successfully achieved in humans. In this research we attempt to design a surface tuned to the selective adhesion of human blood outgrowth endothelial cells (HBOECs). When such a surface is exposed to blood it will scavenge these circulating adult stem cells and the cells will spread, divide, differentiate, and eventually lead to a confluent and functioning endothelial cell layer.

Methods: The polymer system used in this research is a random acrylic terpolymer polymerized from hexyl methacrylate (HMA), methyl methacrylate (MMA), and methacrylic acid (MAA). Acrylic polymers were chosen for their ease of synthesis and biocompatibility. By controlling the molar ratio of HMA, MMA, and MAA we can tailor the physical properties of the polymer [1]. Furthermore, the material has been shown to resist hydrolytic and oxidative degradation *in vitro* [2]. Selective adhesion of the HBOECs is achieved through the covalent incorporation of HBOEC specific ligands found through phage display screening [3]. The ligands were covalently immobilized to the biomaterial through chain transfer chemistry: a novel approach for biofunctionalization [4]. To maximize the cellular response the polymer material was electrospun into a fibrous structure with high porosity. The HBOECs used in this study are obtained by collecting endothelial progenitor cells from fresh human whole blood via density centrifugation and expanded for 4 weeks on rat tail collagen.

Results/Discussion: The first step in this research was a study of HUVEC and HBOEC adhesion to unfunctionalized polymer surfaces (solution cast, random electrospun mats, and aligned electrospun mats). Tissue culture polystyrene (TCPS) was used as a positive control. The HBOEC results from this study are shown in Figure 1 and serve as baseline data to which we can compare HBOEC adhesion to the ligand-derivitized material [5].

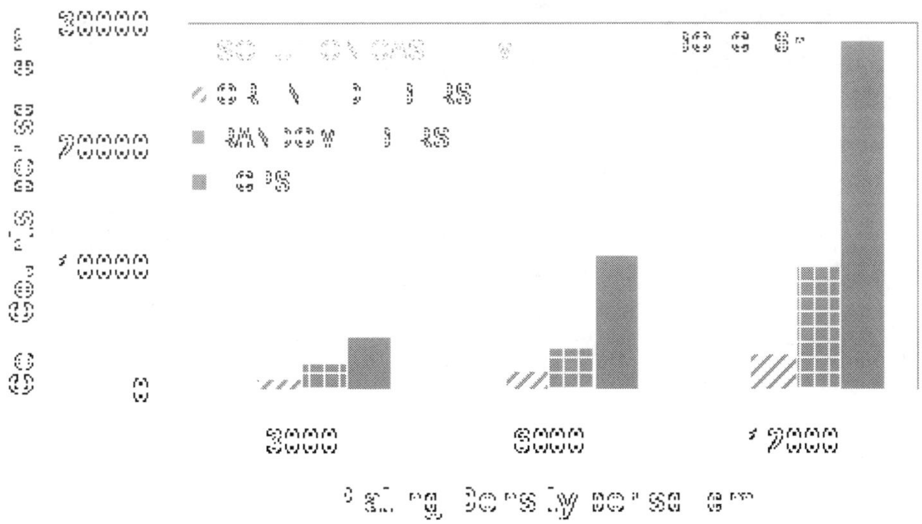

Figure 1: HBOEC adhesion to unfuctionalized polymer surfaces after 6 hours of incubation.

From the phage display study, three ligands were chosen for further exploration (one is a negative control) and were terminated by a GGGSC spacer group whereby the cysteine residue, containing a mercaptan functional group, enables the novel incorporation method. All three of the HBOEC ligands have been successfully immobilized to the polymer and electrospun. An electron micrograph of electrospun functionalized polymer is shown below in Figure 2.

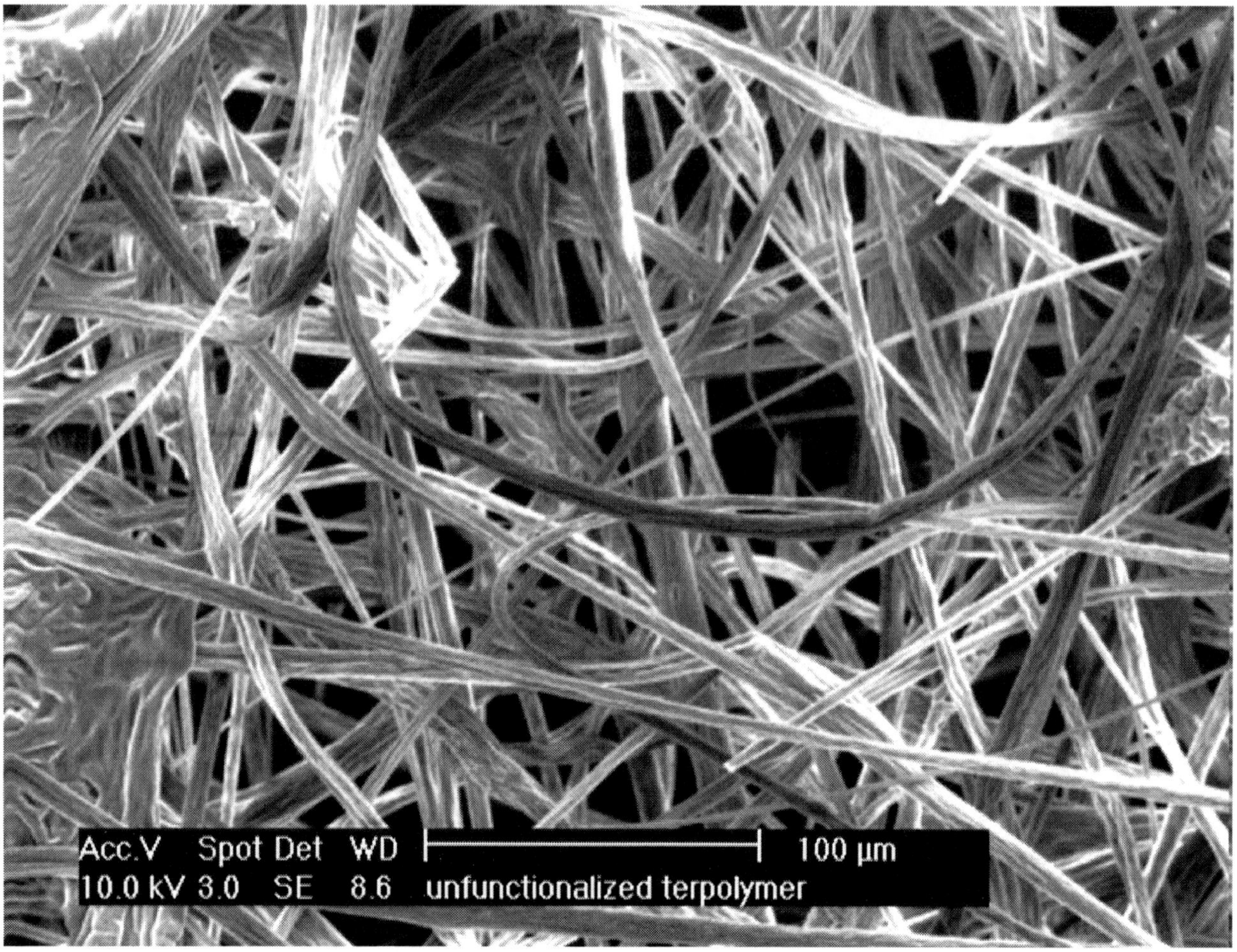

Figure 2: Electrospun biofunctional polymer.

Figure 3 compares the ligand concentration in the polymer before and after electrospinning for selected reactions. The data indicates that the ligands were not degraded during the electrospinning process.

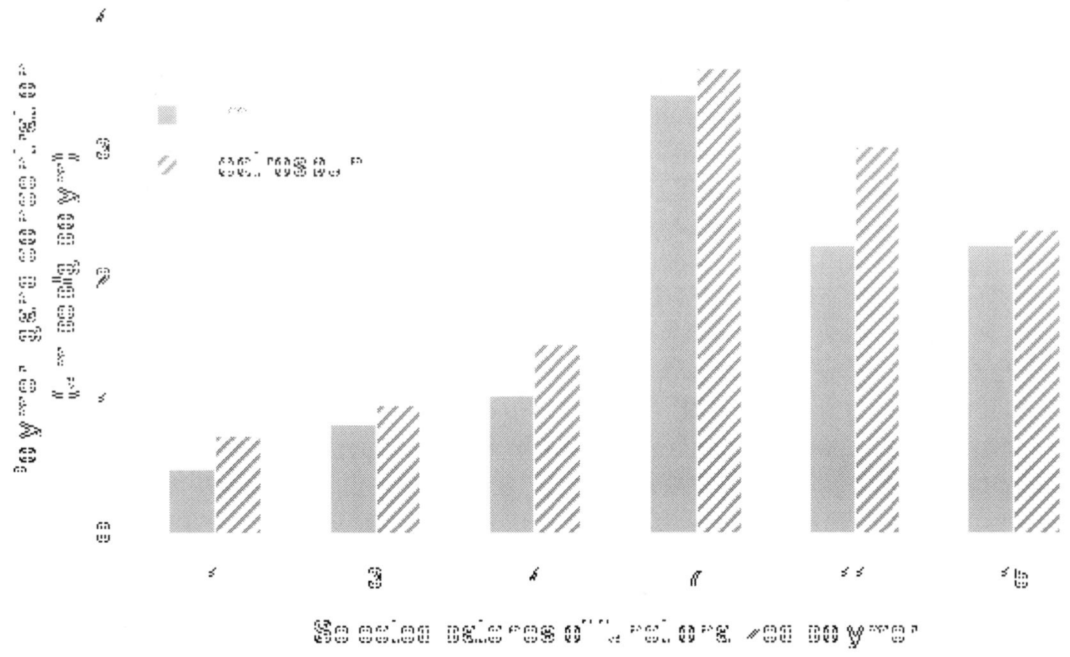

Figure 3: Comparison of ligand concentration of functionalized polymer materials before and after electrospinning.

Data on the HBOEC adhesion and proliferation on ligand containing films and electrospun surfaces will be presented.

Conclusions: The research presented in this abstract illustrates the cytocompatibility of the methacrylate terpolymer base material. Furthermore, we have illustrated the ability to covalently incorporate novel peptide ligands into the polymer material and process this material into fibrous constructs via electrospinning.

References:

[1] Fussell GW, et al, Biomaterials. 2004; 25: 2971 – 2978.

[2] Veleva AN, et al, JBMR. 2005; 74A: 117 – 123.

[3] Veleva AN, et al, Biotechnology and Bioengineering. In press.

[4] Fussell GW, et al, JBMR. 2004; 70A: 265 – 273.

[5] Veleva AN, et al, JBMR. In preparation.

637e Toward Vaults As Drug Delivery Vehicles: Synthesize Encapsulated Polymer Using Entrapped Horseradish Peroxidase

Marcella Yu, Chemical and Biomolecular Engineering Department, University of California, Los Angeles, 5531 Boelter Hall, 420 Westwood Plaza, Los Angeles, CA 90095, Valerie A. Kickhoefer, Department of Biological Chemistry, University of California, Los Angeles, Los Angeles, CA 90095, Leonard H. Rome, Department of Biological Chemistry and California NanoSystems Institute, University of California, Los Angeles, Los Angeles, CA 90095, and Harold G. Monbouquette, Chemical and Biomolecular Engineering Department and California NanoSystems Institute, University of California, Los Angeles, 5531 Boelter Hall, 420 Westwood Plaza, Los Angeles, CA 90095.

Native vaults (72.5 × 41 × 41 nm) are self-assembled, ribonucleoprotein nanocapsules, that consist of multiple copies of 3 proteins (MVP, VPARP, and TEP1) and a small untranslated RNA. Although vaults exist in most eukaryotic cells, their function remains unknown. However, this naturally occurring "organelle" has significant potential as a drug delivery system due to its biocompatibility, large and accessible lumen, and ability to be taken up by mammalian cell lines. The 96 copies of major vault protein (MVP) account for about 70% by mass of the entire vault particle, and stable vault particles can self-assemble from recombinant MVP alone in solution. Our goal is to discover methods to control vault conformational change and structural stability in solution, and to successfully encapsulate material, thereby enabling application of these nanocapsules in drug delivery and nanomaterials synthesis.

Previous studies suggest that the vault has a leaky dynamic structure in solution, which must be overcome if these nanocapsules are to be useful drug delivery vehicles. To resolve this issue, we are investigating the use of an internally synthesized polymer to sequester molecules of interest in a stable state for subsequent delivery. In order to synthesize an encapsulated polymer, horseradish peroxidase (HRP), which catalyzes the polymerization of tyramine, was fused to a protein (INT) that is targeted specifically to the vault interior. The fusion protein (HRP-INT) was expressed using a baculovirus expression system in Sf9 insect cells and then packaged inside of vaults at the crude extract stage. The HRP-INT protein was found to co-purify with vaults and to retain HRP activity, which suggests that the HRP harbored inside the vault remains functional. Current studies include efforts to increase protein expression, and to analyze polymer synthesis using vaults containing entrapped HRP activity.

637f Corneal Epithelial Cell Response To Nano- And Sub-Micron Porous Surface Topographies

Christina Hajicharalambous[1], Xiaoxia Sheng[2], Magdalena Swierczewska[1], Michael F. Rubner[2], and **Padma Rajagopalan**[3]. (1) Bioengineering Program, Lehigh University, Bethlehem, PA 18015, (2) Department of Materials Science and Engineering, Massachusetts Institute of Technology, Cambridge, MA 02139, (3) Chemical Engineering, Virginia Tech, Blacksburg, VA 24061

Corneal opacity is a leading cause of blindness. It can be caused due to several complications such as corneal dryness, diabetes, and injuries sustained by the cornea. The function of corneal epithelium, one of the three major layers in the cornea, is to protect the inner regions of the eye. Corneal epithelial cells present a physical barrier to external agents such microorganisms, and harmful UV radiation thereby protecting the interior regions of the eye, furthermore, the epithelium plays a significant role during wound healing. In vivo, the corneal epithelium is adhered to a basal membrane that exhibits porosity on the nanoscale. We report a simple and versatile approach to fabricate basal membrane mimics that recapitulate the porosity and topography in vivo. Porous membranes were prepared by the assembly of polyelectrolyte multilayers comprised of anionic poly-acrylic acid and cationic poly-allylamine hydrochloride. Porosity on the nano and micro scale was induced by the protonation of carboxylic acid groups and the subsequent reorganization of the polymers upon immersing the films in solutions whose pH ranged from 1.8-3. The porosity can be tuned and controlled to the desired feature size. The average pore diameter measured by AFM was found to be ~100nm and ~ 600nm for the nanoporous and sub-micron porous films respectively. Immortalized corneal epithelial cells were cultured on the porous substrata. Corneal epithelial cell proliferation and migration speeds were significantly higher on nanoporous films. Corneal epithelial cells cultured on nanoporous surfaces exhibit multiple vinculin focal adhesions and well-defined actin cytoskeletal organization. Polyelectrolyte multilayer surfaces with porosity on the nanoscale offer a new route for the growth of corneal epithelium in vitro by mimicking the nanoscale surface topography of basal membranes found in vivo.

639b Nanoscale Zinc Oxide-Enhanced Fluorescence Detection Of Protein Interactions

Nitin Kumar, Adam R. Dorfman, and Jong-In Hahm. Department of Chemical Engineering, Pennsylvania State University, 25 Fenske Building, University Park, PA 16802

Recent advancements in the nanosciences offer a variety of low dimensional materials with precisely controlled and extremely attractive properties which range from improved mechanical strength to unique optical quality. In this Article, we exploit the unique optical properties of nanoscale ZnO materials in the enhanced fluorescence detection of interacting proteins using model biosystems including immunoglobulin G, bovine serum albumin, human serum albumin, streptavidin, and fibronectin. We also devise facile synthetic methods for direct growth of these ZnO nanomaterials in an array format in order to demonstrate their significant potential to serve in applications such as protein or peptide arrays. We demonstrate that the arrayed ZnO nanomaterials function as excellent fluorescence-enhancing substrates in the optical detection of interacting protein molecules without the need for any amplification. Benefited by the outstanding fluorescence-enhancing capability of nanoscale ZnO and its substantial potential for easy integration into biosensor arrays, our new approach may facilitate rapid, high-throughput, ultrasensitive, and specific protein detection that is much needed in the areas of basic biology and biomedical research as well as large-scale clinical testing and screening.

639c Zinc Selenide Quantum Dots As Fluorescent Labels For DNA Detection Applications

Jun Wang[1], Bing C. Mei[1], Tracy Heckler[1], Qi (Grace) Qiu[1], Pedro Lei[2], Stelios Andreadis[2], and T. J. Mountziaris[1]. (1) Chemical Engineering, University of Massachusetts, Amherst, MA 01003, (2) Chemical and Biological Engineering, SUNY-Buffalo, Buffalo, NY 14260

Fluorescent labeling of biological molecules is a technique used widely for analytical purposes in biotechnology and bioengineering. It typically involves the use of an organic dye linked to a moiety that selectively bonds a particular biological molecule, allowing the detection of the latter by the fluorescence of the dye. Typical organic fluorescent dyes have narrow excitation spectra and broad emission spectra that exhibit red tailing.

Semiconductor nanocrystals or quantum dots have emerged as a new class of fluorescent markers with distinct advantages over the traditional organic dyes [1-5]. Their attractive properties include a narrow, symmetric, and strong emission that is size-tunable, continuous excitation by any wavelength smaller than the emission wavelength, resistance to photobleaching, as well as excellent optical and chemical stability that allows their use in lengthy experiments, both in vitro and in vivo. The ability to synthesize different populations of quantum dots with narrow emission spectra permits multiplexing, a property that is very important for simultaneous detection of several analytes, that would be very tedious and expensive if done sequentially.

The focus of this work is the development of new biological detection and DNA analysis schemes by functionalizing the surface of ZnSe-based quantum dots. ZnSe emits in the blue part of the visible spectrum and QDots of this material exhibit size-tunable emission in the UV, violet and blue parts of the spectrum (at wavelengths between 360 and 460nm).

ZnSe quantum dots were synthesized by using two techniques:

(1) Injection of precursors in a hot coordinating solvent [6]. The synthesis of ZnSe quantum dots is carried out in a stirred batch reactor containing liquid hexadecylamine at 260 degrees C. The precursors are diethylzinc diluted in heptane and selenium powder dispersed in trioctylphosphine. The mixture of reactants is injected into the batch reactor and the time of reaction is used to control the size and emission wavelength of the quantum dots.

(2) Synthesis using a microemulsion as template [7-9]. The synthesis was carried out by dissolving zinc acetate in the aqueous dispersed phase of a microemulsion consisting of water as the dispersed phase and p-xylene as the continuous phase with a tri-block copolymer being the surfactant. Hydorgen selenide gas was bubbled through the microemulsion and converted the zinc acetate to ZnSe. By adding acetate salts of Mn and Cu to the aqueous phase we synthesized (Zn,Mn)Se and (Zn,Cu)Se quantum dots that emit in the orange and green parts of the spectrum, respectively.

The formation of water-dispersible fluorescent quantum dots was accomplished by using surface ligand exchange reactions with mercapto-carboxylic acids. The resulting ZnSe quantum dots are stable in aqueous solutions and luminescent over a period of several days, thus suitable for biological sensing applications.

Conjugation of water-dispersible ZnSe-based quantum dots with oligonucleotides was found to increase their fluorescence emission intensity up to fivefold. Quantum dots conjugated with longer DNA strands exhibited stronger fluorescence emission intensity than the ones conjugated with shorter strands, up to a limit of about 50 bases. At the same time, the stability of quantum dots in water was improved. Hybridization kinetics of water-dispersible quantum dots functionalized with complementary short oligonucleotides were also investigated. Upon hybridization, a significant increase (up to 3 times) in the fluorescence intensity of the quantum dots was detected accompanied by a measurable red shift (up to 6nm) in the wavelenght of the fluorescence peak. In another series of experiments DNA hybridization was studied using quantum dots functionalized with oligonucleotides and complementary free oligonucleotides in solution. An increase in the emission intensity of the quantum dots was again observed upon hybridization. A series of experiments has also been performed using free oligonucleotides that are not exactly matched to the ones atatched to the quantum dots. A dependence of the fluorescence intensity on the length and location of the hybridized part was observed. Ongoing experiments in our laboratory are focusing on the development of multiplexed assays for the simultaneous detection of multiple DNA probes functionalized with quantum dots of different sizes and composition that emit at different wavelengths. Applications of these phenomena in the development of DNA detection strategies will be discussed.

Substitution of fluorescent proteins by quantum dot labels in ELISA-type immunoassays is also under investigation aiming to improve both the sensitivity and range of detection of such assays. Successful attainment of these goals will have important implications for the use of quantum dots in high-throughput clinical diagnostic applications, such as real time PCR, DNA microarrays, and immunodiagnostics. We will discuss advances in these efforts using fluorescent ZnSe quantum dots.

References

1. M. Bruchez, M. Moronne, P. Gin, S. Weiss and A.P. Alivisatos, Science 281, 2013 (1998).

2. W.C.W. Chan and S. Nie, Science 281, 2016 (1998).

3. M.-Y. Han, X. Gao, J. Z. Su, and S. Nie, Nature Biotechnology 19, 631 (2001).

4. W. C. W. Chan, D. J. Maxwell, X. Gao, R. E. Bailey, M. Han and S. Nie, Current Opinion in Biotechnology, 13, 40 (2002).

5. X. Gao, L. Yang, J. A. Petros, F. F. Marshall, J. W. Simons, L. Chung, and S. Nie, Current Opinion in

Biotechnology, 16, 63, 2004.

6. M. Hines and P. Guyot-Sionnest, J. Phys. Chem. B, 102(19), 3655 (1998).

7. G.N. Karanikolos, et al., Langmuir, 20(3), 550-553 (2004)

8. G.N. Karanikolos, et al., Nanotechnology, 16, 2372-2380 (2005)

9. G.N. Karanikolos, et al., Nanotechnology, 17, 3121-3128 (2006)

639d Label-Free Colorimetric Detection of Matrix Metalloproteinases on Nanoporous Silicon Photonic Crystals

Lizeng Gao and **Di Gao**. Chemical and Petroleum Engineering, University of Pittsburgh, 1249 Benedum Hall, 3700 O' Hara street, Pittsburgh, PA 15261

Matrix metalloproteinases (MMPs) are degradative enzymes that play an important role in tumor invasion and metastasis. Current MMPs detection methods include zymography, ELISA, western blot, and immuno-histochemistry. However, all of these methods are facing enormous challenges to have the high sensitivity and specificity of MMPs detection required for routine clinical detection of tumors. For example, zymography detection is time consuming and cumbersome; ELISA, western blot, and immuno-histochemistry are not able to detect the activity of MMPs. Here we present the fabrication of a MMPs sensor on a nanoporous silicon photonic crystal platform. The sensor is made by spin-coating a thin gelatin-layer on a nanoporous silicon chip. MMPs present in the sample will digest the gelatin layer and induce a significant change of the rugate peak position on the reflection spectrum of the porous silicon, which can be detected by naked eyes when the chip is illuminated with white light. This method can detect both the presence and the activities of the MMPs, which corresponds to the expression level of MMPs in tumor growth and metastasis. Detection of MMPs at a concentration as low as 1 ng/ml has been demonstrated using this method.

Silver-copper nanoparticles platform for metal-enhanced luminescence

Sanchari Chowdhury[1], Makoto Hirai[2], Venkat R. Bhethanabotla[,1], Ashok Kumar[2] and Rajan Sen[2]*
Department of Chemical Engineering [1], Nanomaterials and Nanomanufacturing Research Center[2], University of South Florida, Tampa, Fl 33620-5350

Introduction

Luminescence based measurements and devices are currently widely used methods in different fields such as biology, chemistry, materials science and medicine. Single molecule detection, DNA sequencing, quantum cryptography, LEDs are the some examples of its numerous diverse applications.

Strong luminescence intensity is one of the most important desired properties for luminophore for its applications. It is possible to design and synthesis luminophores with desired spectral properties. The absorption and emission peaks of any luminophores can be predicted by analogy of known luminophores. But it is difficult to design luminophores with desired luminescence intensity.

Metallic nanoparticles such as silver nanoparticles and gold nanoparticles are known to significantly influence the emission of vicinal luminophores [1,2,4,5,6]. Using nanoparticles platform, it can be possible to increase the quantum yield of weakly luminescent probes by increasing it's emission efficiency by modifying radiative decay rate or by coupling the emission with surface plasmon resonance. It can also improve the intensity of luminophores that already have near unit quantum yield by enhancing their absorption by increasing the local field. Luminescence enhancement ranging from tens to hundred-folds in luminescence signals are reported in literature. A luminescence enhancement phenomenon is strongly dependent on the surface plasmon resonance and the scattering efficiency of nanoparticles. These properties of nanoparticles are dependent on size, geometry and inter-particle distances. Planar metal films are generally known to quench the emission from nearby fluorophores. Distance between the fluorophore molecules and nanostructured platform is also an important parameter in fluorescence enhancement [3]. If the probe molecules are very near to nanoparticles, fluorescence emission from these directly adsorbed onto the surface of metallic nanoparticles is strongly quenched. Similarly, if the probes are too far from the nanoparticles platform effects of these get diminished. So it is important to optimize the distance between the luminophores and nanoparticles.

Although the surface plasmon resonance spectra is one of the most important factors for metal enhanced luminescence, little research has been devoted to establish the role of plasmon resonance spectra. It is reported in literature that the fluorescence enhancement is optimal when the nanoparticles plasmon resonance is tuned to the emission wavelength of the fluorophores[4,5]. Recently, some theoretical[6] and experimental work[5] has suggested that luminescence enhancement is highest when emission wavelength is red-shifted from the plasmon resonance. To obtain information about the relation between surface plasmon

[*] Author for correspondence at venkat@eng.usf.edu

resonance and luminescence enhancement it is important to tune the spectra of nanoparticles at different wavelengths. Surface plasmon resonance wavelength of metal nanoparticles can be tuned by varying their particle size, height, shape, composition, particle-to-particle distance and surrounding dielectric medium. In this work, silver-copper nanoparticles were synthesized using PLD methods. Surface plasmon resonance wavelengths of these Ag-Cu nanoparticles were tuned in the visible and near infrared region by changing the experimental condition such as annealing temperature[7] and substrate temperature. Thus, valuable information about the relation between spectral overlap and luminescence enhancement can be achieved from this study.

To prepare well-defined anisotropic Ag-Cu nanostructures nanosphere lithography is used. This lithography technique is developed by Van Duyne and co-workers[8]. In this method, the planar substrate is coated with polystyrene nanospheres in such away that the nanospheres self-assemble in a single layer or double layer close-packed array. Then the desired metal or semi-conducting material is deposited through the mask created by close-packed nanospheres. Then, the PS nanospheres are dissolved in solvent to leave the truncated nanostructures. Depending on single layer or double layer of polymer nanosphere mask, triangular or hexagonal nanostructures can be obtained. The size of nanostructures and their inter spacing can be changed by changing the nanosphere size and the amount of deposited material on it.

This work will focus on the enhancement of different fluorophore such as platinum octaethyl porphyrin (PtOEP) (important phosphorescent emitter in OLED technology and widely used for oxygen pressure sensing) Ag-Cu nanoparticles platform. The surface resonance wavelength of the Ag-Cu nanoparticles will be tuned to maximize the luminescence enhancement. The study of fluorescence intensity as a function of the distance between a layer of fluorophores and nanostructured metal surface will also be presented.

Experimental Procedure

We created the single layer polystyrene nanosphere (Interfacial Dynamics Corporation, Portland, OR) mask by spin coating. Before coating, the glass substrates were cleaned in a plasma cleaner for 15 minutes. The nanospheres were obtained from manufacturer as a suspension in water then were diluted using surfactant TritonX-100/methanol (1:400) before spin coating. The dilution factor for 300 nm polystyrene nanospheres was 1:2 and for 400 nm it was 1:1. The spin coating speed was 1500 rpm for both 400 nm and 300 nm nanospheres. After coating, the substrates were dried at 60 ^{0}C for 1 hour.

The Ag or Ag-Cu nanoparticles were deposited by using the pulsed laser deposition (PLD) method by using a KrF excimer laser (Lambda Physic, Inc.,LPX 210i) of 248 nanometer wavelength and 200 mj energy and pulse duration 25 ns [full width at half maxima(FWHM)]. The PLD was performed at a repetition rate of 10 Hz. Ag target was used to prepare Ag nanoparticles and Ag-Cu target was used for Ag-Cu nanoparticles. The distance from the target to substrate was 90 mm. After deposition of metal, the polystyrene nanosphere mask was dissolved in ethanol by sonicating for 2 minutes.
Surface morphology of nanosphere mask and nanostructures was observed by atomic force microscope (AFM) (Digital Instruments, Nanoscope IIIa). An electrical furnace (Lindberg, Blue M) was used for the annealing of the Ag–Cu nanoparticles. An UV- Vis spectrometer (JASCO,

V-530) was used for measuring the light absorption spectra attributed to SPR of nanoparticles. A Leica DMI 4000b Inverted Research Fluorescence Microscope equipped with Leica DFC340 FX CCD Camera was used in this study for fluorescence measurements.

Results and Discussion

Using nanosphere lithography and pulsed laser deposition, we fabricated single-layer periodic particle arrays (Figure 1) of silver copper nanoparticles. We got promising results from preliminary study of fluorescence enhancement of PtOEP by Ag nanoparticles grown at different temperatures by the PLD method. Emission peak of PtOEP is at 647 nm. Absorption spectra of Ag and Ag-Cu nanoparticles grown at different substrate temperature and also annealed at different temperature are shown in Figure 1.[1] As absorption spectrum for silver nanoparticles grown at room temperature overlaps with the emission spectrum of PtOEP dye, we found significant enhancement of fluorescence of PtOEP dye on these Ag nanoparticles (Figure 3b) than the sample with no nanoparticles (Figure 3a). On the other hand, Ag nanoparticles grown at 400 ^{0}C did not show any significant enhancement (Figure 3c) as in this case SPR wavelength spectrum does not overlap the emission spectra of PtOEP.

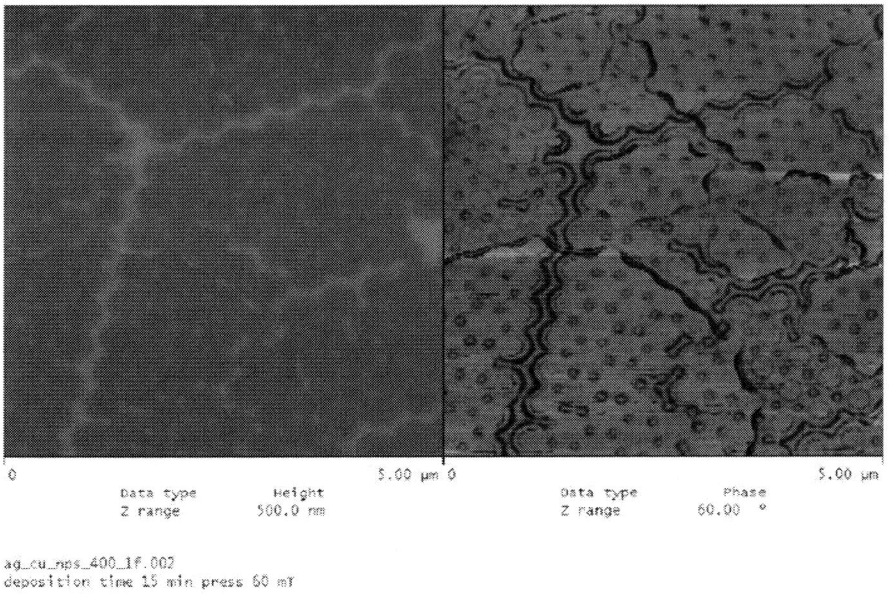

Figure 1: AFM images of silver-copper single-layer periodic particle arrays on glass substrates.

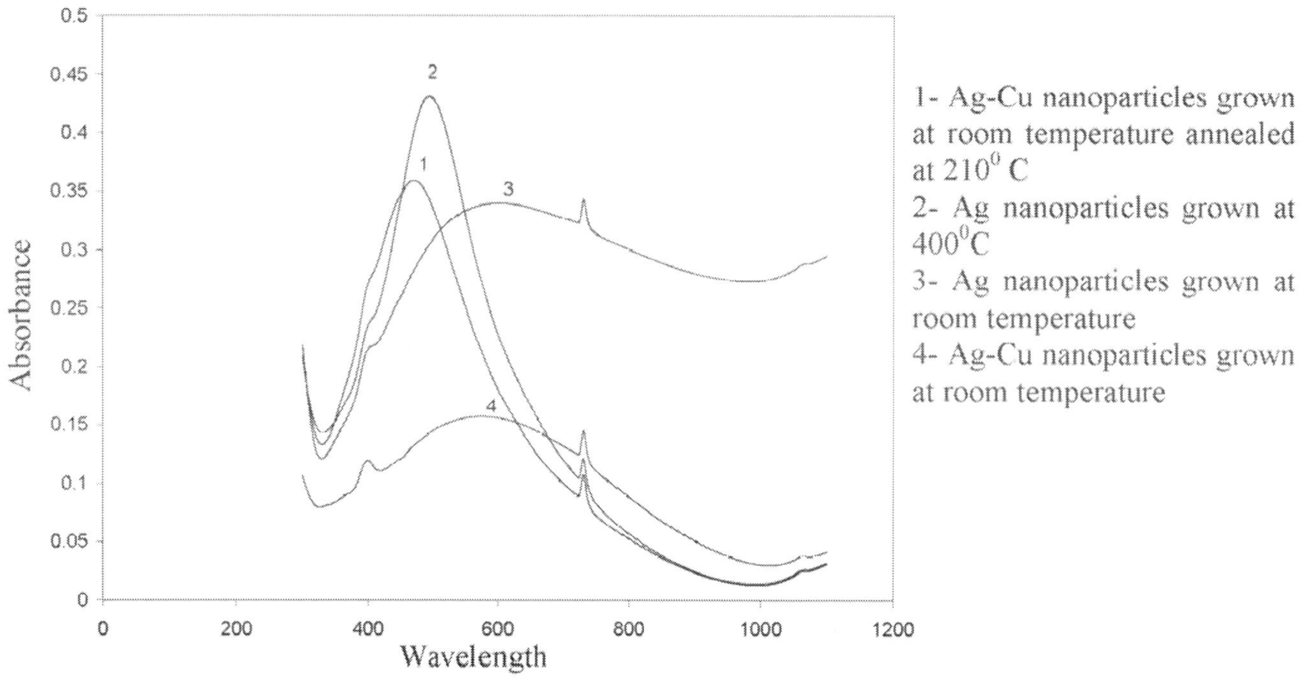

Figure 2: Absorption spectra of Ag and Ag–Cu nanoparticles.

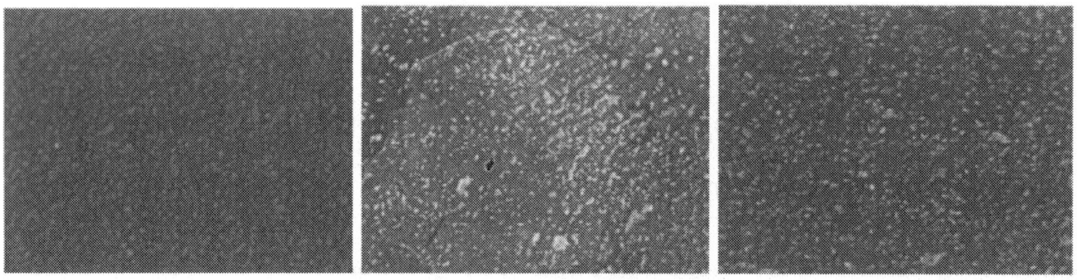

Figure3: Image of PtOEP coated on (a) glass (b) silver grown by PLD at room temperature (c) on silver grown by PLD at 400 C. Images are pseudo colored red to blue indicated lower to higher intensity in arbitrary unit.

Conclusions

In the present work we are attempting to tune the surface plasmon resonance peak of Ag-Cu nanoparticles by changing the annealing temperature and substrate temperature at different wavelengths. We have successfully fabricated single-layer periodic arrays of silver copper nanoparticles. The luminescence intensity of luminophore (PtOEP) at the proximity of Ag-Cu nanoparticles of different SPR spectra is studied to see the effect of spectral overlap on luminescence enhancement. Preliminary result shows that the overlap of surface plasmon resonance spectra of nanoparticles and the emission spectra of luminophore plays an important role in the luminescence enhancement.

References:

1. Lakowicz, J.R., Malicka, J., Gryczynski, I., Gryczynski, Z. Geddes C.D. Radiative decay engineering: the role of photonic mode density in biotechnology. J. Phys. D: Appl. Phys. **36** R240 , (2003)
2. Aslan, K., Lakowicz, J.R., Geddes C.D. Metal-enhanced fluorescence using anisotropic silver nanostructures: critical progress to date. Anal Bioanal Chem 382, 926, (2005)
3. Pan S. Rothberg, L.J. Enhancement of platinum octaethyl porphyrin phosphorescence near nanotextured silver surfaces. J. Am. Chem. Soc. 127, 6087, (2005)
4. Tam, F., Goodrich, G. P., Johnson, B. R. Halas, N. J. Plasmonic enhancement of molecular fluorescence. Nano Lett. 7, 496 (2007).
5. Chen, Y., Munechika, K. Ginger, D. S. Dependence of fluorescence Intensity on the spectral overlap between fluorophores and plasmon resonant single silver nanoparticles. Nano Lett. 7, 690 (2007).
6. Thomas, M., Greffet, J.J. Carminati, R., Arias-Gonzalez, J. R.,Single-molecule spontaneous emission close to absorbing nanostructures, Appl. Phys. Lett., 85 , 3863 (2004)
7. Hirai, M. & Kumar, A. Wavelength tuning of surface plasmon resonance by annealing silver-copper nanoparticles. Appl. Phys. Lett. 100, 014309-1 (2006).
8. Hulteen J C. Van Duyne RP. Nanosphere lithography: A materials general fabrication process for periodic particle array surfaces. J. Vac. Sci. Technol. A 13 (1995)

AUTHOR INDEX

AUTHOR INDEX

AUTHOR INDEX

AUTHOR INDEX

AUTHOR INDEX

AUTHOR INDEX

American Institute of Chemical Engineers
3 Park Avenue
New York, New York 10016-5991

ISBN: 978-1-60423-836-5